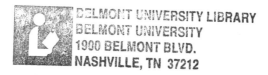

D1218806

Ecology and Conservation of Amphibians

CONSERVATION BIOLOGY SERIES

Series Editors

F.B. Goldsmith
Ecology and Conservation Unit, Department of Biology, University College London, Gower Street, London WC1E 6BT, UK

E. Duffey OBE
Cergne House, Church Street, Wadenhoe, Peterborough PE8 5ST, UK

The aim of this Series is to provide major summaries of important topics in conservation. The books have the following features:

- original material
- readable and attractive format
- authoritative, comprehensive, thorough and well-referenced
- based on ecological science
- designed for specialists, students and naturalists.

In the last twenty years **conservation** has been recognized as one of the most important of all human goals and activities. Since the United Nations Conference on Environment and Development in Rio in June 1992, **biodiversity** has been recognized as a major topic within nature conservation, and each participating country is to prepare its biodiversity strategy. Those scientists preparing these strategies recognize **monitoring** as an essential part of any such strategy. Chapman & Hall have been prominent in publishing key works on monitoring and biodiversity, and with this new Series aim to cover subjects such as conservation management, conservation issues, evaluation of wildlife and biodiversity.

The Series contains texts that are scientific and authoritative and present the reader with precise, reliable and succinct information. Each volume is scientifically based, fully referenced and attractively illustrated. They are readable and appealing to both advanced students and active members of conservation organizations.

Further books for the Series are currently being commissioned and those wishing to contribute, or who wish to know more about the Series, are invited to contact one of the Editors or Chapman & Hall.

Ecology and Conservation of Amphibians

T.J.C. Beebee
School of Biological Sciences
University of Sussex
UK

CHAPMAN & HALL
London · Glasgow · Weinheim · New York · Tokyo · Melbourne · Madras

Published by Chapman & Hall, 2–6 Boundary Row, London SE1 8HN, UK

Chapman & Hall, 2–6 Boundary Row, London SE1 8HN, UK

Blackie Academic & Professional, Wester Cleddens Road, Bishopbriggs, Glasgow G64 2NZ, UK

Chapman & Hall GmbH, Pappelallee 3, 69469 Weinheim, Germany

Chapman & Hall USA, 115 Fifth Avenue, New York, NY 10003, USA

Chapman & Hall Japan, ITP-Japan, Kyowa Building, 3F, 2-2-1 Hirakawacho, Chiyoda-ku, Tokyo 102, Japan

Chapman & Hall Australia, 102 Dodds Street, South Melbourne, Victoria 3205, Australia

Chapman & Hall India, R. Seshadri, 32 Second Main Road, CIT East, Madras 600 035, India

First edition 1996

© T.J.C. Beebee 1996

Typeset in 10/12 Sabon by Saxon Graphics Ltd, Derby
Printed in Great Britain by St Edmundsbury Press, Bury St Edmunds, Suffolk

ISBN 0 412 62410 9

A catalogue record for this book is available from the British Library

Library of Congress Catalog Card Number: 95-71856

∞ Printed on permanent acid-free text paper, manufactured in accordance with ANSI/NISO Z39.48–1992 and ANSI/NISO Z39.48–1984 (Permanence of Paper).

Contents

Preface

Herpetology, the study of amphibians and reptiles, can be a difficult science to sell. I guess that most of its practitioners will recognize my experiences of cynical amusement from friends and colleagues astonished that anyone should spend a lifetime working with amphibians, let alone that government funds can (sometimes) be obtained to support such apparently esoteric research. But the case is strong, and we have no need for defensive posturing; amphibians are ideal subjects for many kinds of ecological studies that can throw new light on fundamental questions in the natural sciences, and at the same time they exemplify the plight of wild creatures suffering declines all over the planet at human hands. Now, more than ever, we need to understand ecosystems and commit the necessary resources to their protection. If we fail to do this on a scale as yet not appreciated by most political leaders, and within the next two or three human generations, the consequences will be dire. So no apologies are in order; amphibians are fascinating and important, and we must look towards a better understanding and protection of them.

In this book I have attempted to bring together current knowledge in all areas of amphibian ecology and conservation. Such an ambition has several consequences. One is that exhaustive coverage of so large a subject is impossible within the space available, and the text is therefore furnished with what I hope is a representative rather than a comprehensive bibliography. Another is that although amphibians occur in all continents of the world except Antarctica, my reporting has been biased away from the more descriptive studies that still dominate the literature of many countries in favour of the experimental approaches that are increasingly dominant in Europe and North America, a disparity of emphasis highlighted by Shine (1994) in his plenary lecture at the Second World Congress of Herpetology in Australia. I must also admit to a deficit in coverage of herpetological studies carried out in the old Soviet Union; although important work has been done there (e.g. Ishchenko, 1989), most of it has unfortunately remained inaccessible to outside audiences.

Innumerable friends and colleagues have stimulated and sustained my personal obsession with amphibians, and it is a pleasure to thank them not only for encouragement and companionship but in several cases also for generous material contributions towards this publication. My special thanks, therefore, to Claes Andren, Henry Arnold, Pim Arntzen, Brian Banks, Rob Beattie, Maggie Beebee, Andrew Blaustein, John Buckley, Sarah Bush, Clive Cummins, Arnold Cooke, Keith Corbett, Jonty Denton, Tony Gent, Paul Gittins, Richard Griffiths, Kurt Grossenbacher, Tim Halliday, Sue Hitchings, R. Krekels, Tom Langton, Herb Macgregor, Rob Oldham, Chris Reading, Paul Rooney, Ulrich Sinsch, Fred Slater, Henk Strijbosch, Ton Stumpel, Mary Swan, Maria Tavano and Adeline Wong; also to David Streeter and Barrie Goldsmith for promoting the idea of this book, and Colin Atherton for photography of figures. In large part it is the work of these people that has made this book possible.

Ecology is of course a scientific enterprise, but for me the real charm of amphibians lies beyond the strictly academic domain. Perhaps for those of us in temperate lands it has something to do with the passing of the seasons and, in particular, the role of frogs and newts as harbingers of the coming spring. The special vitality of the dawning year is surely one of our most heartfelt delights, and amphibians are of that time.

> Spring came haunting my garden today –
> A song of cold flowers was on the grass.
> Tho' I could not see it
> I knew the air was coloured
> And new songs were
>   in the old blackbird's throat.
> The ground trembled at the thought
>   of what was to come!
> It was not my garden today,
>   it belonged to itself.
>
> *Spring Song*, Spike Milligan.

Trevor Beebee 1995

─── 1 ────────────────────

What are amphibians?

1.1 THE CLASS AMPHIBIA

1.1.1 General features

A self-evident prerequisite for anyone setting out to study a particular group of organisms is a familiarity with their basic biology. Inevitably this must include the fundamentals of taxonomy and an understanding of the general life styles that are characteristic of the group in question. In addition, some knowledge of anatomy and physiology is also important for anyone with more than a passing interest in amphibians. The basics of amphibian biology are therefore covered in this chapter, with an emphasis on those aspects of special importance to ecology and conservation; a more complete account is given by Duellman and Trueb (1986).

There are thought to be 4000 or more extant amphibian species, in comparison with a similar number of mammals, more than 20 000 fishes, 8000 reptiles and 9000 birds. The 4000 amphibians are divided, very unequally, into three Orders. These are the Anura (or Salientia), the 'tailless amphibians' or frogs and toads, by far the largest group with some 3500 species; the Urodela (or Caudata), newts and salamanders, with at least 350 species; and the Gymnophiona (Apoda), worm-like, burrowing caecilians with over 150 known species but with which this book is not concerned because their ecology remains mostly unknown.

Common amphibian characteristics include shell-less eggs and a naked, more or less permeable skin richly endowed with secretory glands but without fur, feathers or scales. Most amphibians therefore require high levels of humidity, or a fully aquatic environment, in which to live. There are exceptions to this and all around the world species can be found which have adapted to conditions as dry as any tolerated by apparently more robust organisms; thus the South American leaf frog *Phyllomedusa bicolor* produces a waxy secretion which it spreads all over its body to minimize water loss, several genera of spadefoot toads spend much of their lives in damp burrows below ground in otherwise arid habitats, and a number of highly adapted desert frogs retire underground to produce temporary, impermeable cocoons that tide them over until rains come again. Australian cocoon-making species have long been known to the native Aboriginals as sources of water when dug out of their subterranean retreats.

2   What are amphibians?

Amphibians are probably best known for a mode of reproduction in which eggs are laid in water, hatch into tadpoles and subsequently undergo metamorphosis into a morphologically distinct adult form. It is worth pointing out that this pattern, while long familiar to naturalists in the northern hemisphere, is much less typical of tropical species and thus (because these constitute an overwhelming majority of the global total) not a general feature of the class as a whole. Fertilization may be internal or external but, with the exception of caecilians (which have a phallodeum) and one species of anuran (the 'tailed' frog *Ascaphus truei*), males adopting the former method have no proper intromissive organ and other methods of sperm transfer are employed. In both cases there is usually some form of embrace (refererred to as **amplexus**) between the sexes, though this may be brief and has been done away with altogether in some urodeles (newts).

Despite their name, amphibians are amphibious only in fresh or mildly saline water; no species can survive prolonged exposure to sea-water, though the extent of tolerance to saline conditions does vary considerably between species: the crab-eating frog *Rana cancrivora* thrives in the mangrove swamps of south-east Asia and its larvae can survive at salinities corresponding to 80% sea-water (NaCl about 2.2%, or 0.375 M), a remarkable feat unparalleled by any other known species (Dunson, 1977). Amphibians do not drink, but absorb the water they need directly through their skins during immersion in ponds or streams. Osmoregulation in amphibians exposed to high salt concentrations, or prolonged drought, involves the accumulation of high plasma concentrations of urea; this can reach 0.26 M in European green toads (*Bufo viridis*) and an astonishing 0.35 M in crab-eating frogs (Degani and Hahamou, 1987).

Although amphibians are ectothermic, several genera penetrate the Arctic Circle and cold-adapted species occur in Eurasia and the Americas. Many of these are forced into hibernation for more than half of each calendar year and this may be accomplished deep underground, or underwater often beneath thick ice sheets. Of particular interest are frogs which can survive being deep-frozen in ice, such as the North American tree frog *Hyla crucifer* and wood frog *Rana sylvatica*. These and a few other species accumulate high concentrations of glycerol or glucose (up to 0.5 M), which act as cryoprotectants, in their blood and body tissues (Schmid, 1982; Costanzo and Lee, 1993) but the ability to withstand freezing in ice (as opposed to long periods at 4°C beneath it) is evidently rare in the Amphibia. Some species, especially aquatic salamanders such as *Ambystoma maculatum* in North America and *Proteus anguineus* in Europe, are fully active at temperatures between 0°C and 5°C and urodeles in general tend to operate at lower temperatures than anurans. Thus the range for a series of temperate and tropical American salamanders was from −2°C to 30°C with a mean of 13.9°C, whereas the comparable data for anurans were 3°C to 35.7°C with a mean of 21.7°C (reviewed in Duellman and Trueb,

1986). There is much less evidence of thermoregulation to maintain a constant body temperature in amphibians than there is for reptiles, but basking behaviour by some anurans suggests that attempts at such regulation do occur. Critical maximum temperatures for the most tolerant species (such as the tree frog *Hyla smithii*) are in the range of 38–42°C.

Amphibians have no fewer than four respiratory mechanisms, employed to widely different degrees according to species and habitat. Thus gas exchange can take place through gills or lungs, across the surface of the skin, or in the buccal cavity. The latter method requires that the throat pulsates constantly to ventilate its respiratory surfaces, an obvious feature of most amphibians while living on land.

Other attributes of amphibians include a heart with two atria but a single ventricle, vestigial ribs which do not enclose the thorax, two occipital condyles at the rear of the skull (which interact with processes on the atlas vertebra) and a single sacral vertebra. Modern amphibians are all small in comparison with other aquatic or terrestrial vertebrates; the largest living today are the goliath frog *Conraua goliath* of the Cameroons, with a body length of about 30 cm; the Chinese and Japanese giant salamanders *Andrias davidianus* and *A.japonicus*, which may reach total lengths of 1.6 m; and the Colombian caecilian *Caecilia thompsoni* at about 1.5 m. Most amphibians, however, are very much smaller than these extremes.

1.1.2 Food and feeding

Amphibians are almost without exception carnivorous in the adult state. Vegetable matter is rarely found in amphibian guts, and though some aquatic salamanders (*Siren* spp.) and one or two species of frogs may be partly or wholly vegetarian (Stocks, 1992) the bias towards carnivory is much greater than in any other vertebrate class. A sweeping statement, but generally true, is that anything that is small enough to be swallowed is likely to be taken as prey; the few exceptions include distasteful or toxic species, or those capable of delivering a deterrent bite or sting. The effectiveness of such deterrence varies considerably and unpredictably; thus the tadpoles of bufonid toads are distasteful to most adult amphibians, including the European newts *Triturus vulgaris* and *T. helveticus*, but are consumed avidly by the often sympatric *T. cristatus* (Cooke, 1974). The major prey items of amphibians are invertebrates, especially insects, crustaceans, worms and molluscs, but larger, vertebrate organisms are sometimes taken by the bigger species. There are many accounts of rodents, fledgling birds, fish, other amphibians (including smaller conspecifics) and reptiles (including venomous snakes) featuring on the diet sheet.

Most terrestrial anurans catch their prey by rapid protrusion of a sticky tongue (which is hinged at the front of the mouth rather than the back), often accompanied by a flying leap in the direction of the potential victim. The entire capture event is over in milliseconds and cannot be followed by eye. Aquatic pipids (such as *Xenopus*) lack tongues and simply grab food items by mouth and suck them directly inside. Aquatic salamanders and newts similarly lunge towards prey and seize them directly in the mouth, but many terrestrial urodeles catch their food by tongue-flick rather like anurans do. Most dramatic are some of the lungless plethodontid salamanders, such as *Bolitoglossa occidentalis*, which can project their tongues relatively large distances (more than 50% of their body length), frontally or laterally and with great speed, to catch bees and flies on the wing. Caecilians approach their prey slowly and then make a firm, often lateral, bite of seizure. Many have long, dagger-like teeth to increase the efficiency of this process though their prey are usually earthworms and thus slow-moving and readily caught.

1.1.3 Defence

Amphibians are, by and large, poorly endowed with tooth and claw. It is true that the larger caecilians and some salamanders (such as *Amphiuma* spp.) have sufficient and sharp enough teeth to administer a memorable bite, and a few of the biggest anurans (such as the African bullfrog *Pyxicephalus adspersus* and South American *Ceratophrys* toads) are similarly capable, but in most species these weapons are small or, as in bufonid toads, completely absent. More usual adaptations include a heavy reliance, through cryptic coloration, on not being detected in the first place; rapid flight by leaping or powerful swimming, or both, sometimes accompanied by alarm calls; feigning death; and the adoption of characteristic defence postures, including the so-called 'unken reflex', named after the sound made by one of its practitioners, the European fire-bellied toad *Bombina bombina*. Simple defence postures involve hyper-inflation with air and body tilting to present a larger-than-life image to would-be predators such as snakes that can be influenced by their perception of prey size. The unken reflex, on the other hand, is a stationary display adopted by anurans and urodeles bearing aposematic coloration, in an attempt to show their warning signals to the best advantage. Some urodeles have developed rather specific, sometimes bizarre ways of protecting themselves. The Iberian ribbed salamander *Pleurodeles waltl* has ribs which, when its body is grasped, penetrate the skin and presumably the would-be predator as well; and a few species, such as the European gold-striped salamander *Chioglossa lusitanica*, can readily autotomize their tails in much the same way that many lizards do.

Arguably the most important amphibian defence system is the armoury of bioactive chemicals that many species produce in specialized skin glands. As with several other amphibian features, these noxious and toxic substances are at their most primitive (though also rather little studied) in the caecilians, more advanced in the urodeles and at their most sophisticated in the anura. Salamanders in the *Salamandra* genus, for example, produce toxic alkaloids from the large parotid glands on the neck when threatened or roughly handled, and various newts of the *Taricha*, *Notophthalmus* and *Triturus* genera produce a range of obnoxious neurotoxins and poisonous proteins under similar circumstances. Human tasters of these skin secretions have provided first-hand accounts of their discomforting effects on mammalian physiology (e.g. Ormerod, 1872). Among the anura, bufonid toads synthesize a variety of cardiotoxic steroids including bufotoxin and bufogenin, as well as potent hallucinogens such as O-methyl bufotenin that have engendered dangerous habits, such as toad-licking, in some countries recently.

By far the most potent chemical defences, however, are those possessed by the dendrobatid 'arrow poison' frogs of Central and South America. These animals synthesize a range of alkaloids based on the simple piperidine 6-membered ring structure, or in the case of the *Phyllobates* group, the extremely toxic alkaloid batrachotoxins. The latter are among the most dangerous non-protein toxins known, and the material from a single specimen of the bright yellow *P. terribilis* would suffice to kill perhaps 10 people. As their name suggests, the secretions from these frogs have long been used to smear arrows used for hunting by the native peoples of the American tropics. No doubt these poisons are effective deterrents to most potential predators, and dendrobatid frogs have extraordinarily bright aposematic coloration as well as bold, unsecretive behaviour. They are of course immune to their own toxins, but are not entirely safe and it seems that at least one snake (*Leimadophis epinephelus*) is able to devour them with impunity (Mattison, 1987). Other amphibian toxins give substantially less protection. For example, European bufonid toads are successfully predated by several species of mammal, bird and reptile, and in Britain specimens of *Bufo bufo* have even been recovered from the stomachs of migratory fish (Falkus, 1977).

Toxin production can start early in development, particularly in anurans. In many species of *Bufo*, tadpoles and even eggs have sufficient quantities to deter many would-be consumers.

Chemical defences are not just directed against predators but also at the vast armada of pathogenic microbes for which the moist, soft skins of many amphibians would otherwise provide an ideal habitat. Small peptides isolated from frog skin, such as magainin, show considerable clinical promise as antimicrobial agents active against gram-negative and gram-positive bacteria and against fungi (Gabay, 1994).

1.1.4 Sensory systems and learning abilities

Eyes are well developed in anurans and most urodeles but degenerate in some fossorial salamanders and burrowing caecilians. Rods and cones are present and it seems likely that many amphibians can discriminate a wide range of colours. Despite this, and the flamboyant coloration of many amphibian species, there is little evidence that colour discrimination is widely used in practice.

Vision is critical in prey capture for most anurans and many urodeles, and is particularly keen for detecting movement. Although each species has its particular optimum light intensity for visual acuity, many are especially adept at low light levels. The European common toad *Bufo bufo*, for example, can catch prey efficiently using visual cues at light intensities as low as 10 μlux – conditions similar to those at midnight in dense woodland (Larsen and Pedersen, 1982). Many amphibians can detect prey by sight over distances of more than 1 m, and larger moving objects (potential predators) generate a response when still tens of metres away. Vision is also important in orientation, and amphibians can respond to polarized light even when the sun is obscured by cloud cover.

Amphibians are equipped with sensitive auditory receptors and respond to a wide range of sounds. Ears are especially well developed in anurans, where they feature strongly in breeding behaviour. Hearing also plays a part in prey and predator detection in some species, and all amphibians are sensitive to seismic vibrations (10–500 Hz) through the substrate, detecting them via the sacculus and lagena of the inner ear. It is probably this sense which alerts, and shuts down, the nocturnal breeding choruses of some anurans in response to the footsteps of approaching human visitors long before they are close enough to be seen.

Olfaction is well developed in all classes of the Amphibia, though the sensory apparatus is at its most sophisticated in anurans. The sense of smell serves at least two purposes: it is concerned with prey detection and with orientation. In the case of caecilians and some urodeles, olfaction is the primary method for finding food. Newts kept in aquaria will seek out non-motile prey by scent alone, especially if the victim has already suffered tissue damage, while caecilians in their subterranean burrows have only olfactory and auditory cues available to them in their hunt for worms. The extent to which anurans use olfaction to find their prey remains uncertain; experimental evidence shows that olfaction can be used by at least some species, especially aquatic frogs such as *Xenopus*, but visual cues seem to be much more important for most terrestrial frogs and toads. On the other hand, it is clear that both anurans and urodeles use olfaction as a major sense when orienting either within their home ranges or towards breeding sites. The classic studies with European frogs *Rana temporaria* (Savage, 1961) and with the North American newt *Taricha rivularis* (Twitty, Grant and Anderson, 1967) demonstrated unequivocally the importance of olfaction in this area and are among the seminal works of modern herpetology.

Other senses are also important to amphibians in various ways. Tactile cues can be used when animals are close enough to grab one another, often to identify sex or conspecificity. Thus it is possible by touch alone for humans (and therefore presumably other frogs) to identify female *Rana temporaria* because of the abundance of pearly granules on the abdominal skin, which in males is completely smooth. Tadpoles and some adult amphibians, especially highly aquatic ones, also have 'lateral line' detector organs along the sides of the trunk and on the head, similar to those widely occurring in fish. They are sensitive to water currents, and in newts probably play a role in breeding behaviour during displays in which males create such currents and direct them towards females by tail action. Finally, at least some species of anurans and urodeles (including bufonid toads, plethodontid cave salamanders and the newt *Notophthalmus viridiscens*) can sense and respond to variations in the earth's magnetic field, and use this ability for orientation purposes (Adler, 1982; Sinsch, 1991).

Amphibians have limited but not negligible abilities to learn from their experiences. Although they do poorly in artificial mazes, it is becoming clear that many terrestrial species do generate some kind of memorized image of their spatial surroundings, and in this way become familiar with their home range topography. The observations of some amphibians migrating to sites of breeding ponds that have been totally obliterated by development also imply the existence of a long-term memory. It is well known that in captivity toads of the genus *Bufo* can learn to avoid seizing stinging insects such as wasps, and that once learnt the lesson is remembered for several days or even weeks. There is no doubt that in captivity many amphibians become familiar with humans to the extent that, when their captor appears, the flight response is suppressed by one of food expectation. Toads may even chase people in this situation.

1.1.5 Reproductive strategies

Yet another set of superlatives applies to amphibian reproduction: no other class of vertebrates matches them in the sheer variety of ways in which they procreate. In temperate countries, most anuran species experience an annual cycle of reproductive activity which may have a single, sharp peak (so-called 'explosive' breeders) or a more protracted one in which females sometimes lay several egg clutches (Wells, 1977). A substantial number of salamanders (especially the terrestrial plenthodontids) and some anurans (such as *Rana pretiosa* in mountainous areas of the United States) regularly reproduce biennially, however. By contrast, tropical species of all three amphibian classes usually breed throughout the year, often in response to extrinsic factors such as heavy rainfall.

In urodeles, fertilization is usually internal but there are exceptions, with males of the entirely aquatic Asiatic salamanders (Hynobiidae), giant salamanders (Cryptobranchidae) and sirens (Sirenidae) shedding sperm over eggs after they are deposited, much in the way of most anurans. Some urodeles with internal fertilization lay their eggs in water and the larvae develop in the same medium; European newts of the genus *Triturus* are classic examples. Others lay their eggs on land, and these hatch into aquatic larvae after flooding of the nest site (as with some *Ambystoma* species) or may develop completely in the nest without ever entering water (as with the plethodontid *Desmognathus aeneus*). In yet other species the eggs are retained in the oviducts and then deposited in water as fully formed aquatic larvae (the usual situation in the common European *Salamandra salamandra*) or even retained within the mother until metamorphosis, as with the alpine *Salamandra atra* which gives birth to fully formed juveniles. Fecundity is also very variable; it is highest in large species, as might be expected, and also in those that deposit eggs in water. Thus the tiger salamander *Ambystoma tigrinum*, can produce more than 5000 eggs in one season, while many newt and salamander species lay just a few hundred eggs each year. At the other extreme, live-bearing terrestrial species such as *Salamandra atra* produce just one or two young. Parental care also occurs among urodeles; either sex of some fully aquatic salamanders (such as the European olm, *Proteus anguineus*) attends the eggs and appears both to defend and to oxygenate them, and females of some species that oviposit in terrestrial nests (such as *Ambystoma opacum*) remain there until hatch-time. Among the terrestrial plethodontid salamanders, nest attendance is virtually ubiquitous (Figure 1.1). These parents defend the eggs against potential predators, agitate them to improve oxygen supply and remove dead ones that become infected with fungus.

Figure 1.1 Female cave salamander *Speleomantes ambrosii* brooding eggs. (Photo: M. Tavano. Reproduced, with permission, from Salvidio *et al.* (1994), *Amphibia–Reptilia* **15**, 35–46, published by E.J. Brill.)

In anurans, unlike the other amphibian orders, external fertilization is the norm although there are a few exceptions among tropical species such as *Eleutherodactylus coqui* (Townsend *et al.*, 1981). The commonest mode of anuran reproduction follows the classic pattern of egg deposition and larval development in lentic or lotic freshwaters, but there are in this case many variations on the theme as well as exceptions to the rule. An extreme version of the classic mode, used by a considerable number of frog species, is the deposition of eggs in the small water volumes of tropical bromeliads; another is the production of eggs with sufficient yolk to see the larvae through to metamorphosis without any need to feed, as achieved by a range of anurans in South America, Indonesia and Madagascar. Yet another group of frogs lays eggs on land which hatch into larvae that fall or crawl into nearby freshwater; and numerous tropical species make foam nests in which tadpoles hatch either in trees overhanging water or on a bank next to it, and from which they gain access to the water for normal larval development. As with urodeles, some anurans have forsaken the tadpole phase altogether, laying eggs on land or in vegetation which give rise not to tadpoles, but to fully formed froglets. By contrast, complete development within the mother's oviducts is rare, but it does occur in a few species and may be primitive (with nutrients provided by yolk) or advanced with the provision of nutritious oviducal secretions, as in *Nectophrynoides occidentalis*. In general, the greatest variety of anuran reproductive modes occurs in the tropics, while at high or low latitudes the great majority of species adopt the classical strategy popularized in so many books on amphibian biology.

Fecundity in the anura covers an even greater range than in the urodela but follows a similar pattern. Once again large, water-breeding species produce the most eggs, with the American bullfrog *Rana catesbiana* and the marine toad *Bufo marinus* regularly depositing more than 20 000 at a single sitting. At the other end of the spectrum, the tiny terrestrial Cuban frog *Sminthillus limbatus* lays a complete clutch of just one egg, which ultimately produces a fully formed froglet.

Anurans exhibit a greater range of parental care behaviour than either the caecilians or the urodeles. About 10% of known anuran species exhibit some form of care and this can be performed by males or females, according to species; the method varies from the relatively rare protection of aquatic eggs and tadpoles, such as the driving off of potential avian predators by female *Leptodactylus ocellatus* and the remarkable channel construction by male African bullfrogs (*Pyxicephalus adspersus*) to rescue desiccating tadpoles, through to the rather commoner attendance at terrestrial nests exhibited by, for example, certain leptodactylids, ranids and microhylids. Rather more sophisticated are those anurans which carry eggs or tadpoles around with them; among aquatic species this occurs in Surinam toads (*Pipa* spp.), in which eggs develop into tadpoles while implanted in the mother's back and then escape into the water, but as a stratagem it is much more frequent among the terrestrial anura. Male European midwife toads (*Alytes* spp.) carry eggs entwined around their legs and release tadpoles into ponds when the eggs hatch; tropical *Gastrotheca* frogs, among others, carry eggs in a dorsal pouch until hatch time, while dendrobatid frogs carry tadpoles adhering to their dorsal surface; male Darwin frogs (*Rhinoderma* spp.) transport tadpoles in their vocal sacs. In the truly exceptional (and possibly now extinct) gastric brooding frogs (*Rheobatrachus* spp.) of Australia, the eggs are swallowed by the female; tadpoles develop in her stomach during an enforced fast and are finally regurgitated after metamorphosis as fully formed froglets. Perhaps equally remarkable is the fact that some frogs even feed their young; female *Dendrobates pumilio*, among others, deposit their tadpoles in bromeliad pools and then revisit periodically to deposit unfertilized eggs there which the tadpoles eat.

1.1.6 Genetics

Amphibians have not been widely used for classical genetic studies, perhaps because they lack the short generation times of insects such as *Drosophila* and the relevance to humans of mammals like laboratory mice. However, by way of interesting examples, Spurway (1953) carried out mating experiments between the four species of closely related European crested newts *Triturus cristatus*, *T. karelinii*, *T. dobrogicus* and *T. carnifex*, and was able to show that many of the small differences used to distinguish them (all were considered at the time to be subspecies rather than species) segregated in Mendelian fashion and are thus each determined by a single, unlinked genetic locus.   Cross-breeding experiments with several species exhibiting coloured vertebral stripes (such as the frog *Rana limnocharis* and the salamander *Plethodon cinereus*) have shown that in all cases presence of the stripe is dictated by a single dominant allele. Albinism, fairly common in the amphibia, is also a single-locus mutation (this time recessive) in *Xenopus laevis* (Hoperskaya, 1975).

Sex determination in amphibians is not uniform. In most species there are no gross size differences between the sex chromosomes, so sex determination by karyotyping (which would be very useful in larvae) is not possible. There are exceptions to this rule and other cases where sex chromosomes can be demonstrated by band-staining procedures. Examples of species in which karyotyping can be used for sex determination include the anurans *Pyxicephalus adspersus* and *Rana esculenta*, newts (several *Triturus* species) and the salamander *Necturus maculosus*. There is also variation with respect to which is the heterogametic sex; thus in *Rana*, and the newts *Triturus alpestris*, *T. vulgaris* and *T. helveticus*, the male is heterogametic; whereas in *Xenopus* and other species of *Triturus* (*cristatus* and *marmoratus*) it is the female.

Polyploidy occurs occasionally in numerous amphibian species, and triploid all-female populations of *Ambystoma* salamanders occur in the Great Lakes region of North America. Mostly, these are not true parthenogens because they require mating with males of sympatric diploid species to produce viable eggs; the role of the sperm, however, is purely as a physical initiator of embryogenesis and not as a donor of genetic information. On at least one island with a population of triploid females, no male ambystomas have yet been found, and thus true parthenogenesis may be in operation.

Finally, environmental sex determination may have a role in the Amphibia as it clearly does, on a much larger scale, in the Reptilia (Gutzke, 1987). In the newt *Pleurodeles waltl*, in which females are the heterogametic sex, rearing larvae at approximately ambient temperatures (around 20°C) produces a normal 1:1 sex ratio, but elevating larvae to 30–31°C during a critical phase of development causes complete conversion of genetic females into phenotypic males (Dournon and Houillon, 1985). The ecological significance of this observation, and its relevance to other amphibian species, remains to be determined.

Polyploidy has sometimes been a feature of speciation in amphibians, a classical example being the morphologically very similar North American tree frogs *Hyla chrysocelis* (diploid with 24 chromosomes) and *H. versicolor* (tetraploid with 48 chromosomes). These species are sympatric, and most easily distinguished by differences in male mating calls.

Table 1.1 DNA contents of amphibian cells

Order/family	Number of species examined	DNA (pg) per cell (range or average)
Gymnophiona	3	7–28[a]
Urodela		33–192[a]
Plethodontidae	9	57
Salamandiidae	5	54
Proteidae	1	165
Ambystomatidae	1	48
Cryptobranchidae	1	112
Hynobidae	3	37
Sirenidae	1	108
Amphiumidae	2	156
Anura		2–36[a]
Discoglossidae	3	17
Bufonidae	1	10
Hylidae	2	6
Pelobatidae	2	9
Ranidae	10	11
Pipidae	2	6

[a] Total range of C-values for Order

The sheer size of amphibian chromosomes is a distinctive feature of the group. Those of the great crested newt *Triturus cristatus*, for example, are some 10 times larger than their human equivalents (Figure 1.2). This species is bizarre because viability requires heterozygosity in chromosome 1, with the consequence that 50% of zygotes always die as a result of this enormous genetic load (Macgregor and Horner, 1980). It turns out that crested newts also have much more DNA (about sixfold) than humans, though there is no reason to believe that newts have more genes than people do. Amphibians represent an extreme example of the so-called **C-value paradox** – the observation that all higher organisms, unlike bacteria and viruses, have far more DNA (the total amount of which in the haploid genome represents the organism's C-value) than is required to make up the complement of genes. Thus whereas human somatic cells, which are fairly typical of mammals, each contain about 7 picograms of DNA, most caecilians, frogs and toads have as much or more while newts and salamanders usually have a lot more (Table 1.1). The mud-puppy (*Necturus*) and eel-like *Amphiuma* salamanders of the United States are the most extreme examples known, with some containing upwards of 180 pg. Only lungfishes and lilies are on a par with these amphibians in this curious property.

Why do amphibians have so much 'junk' DNA? One interesting observation is that correlations exist between DNA content, individual cell size and the larval development period. A hypothesis which takes account of these facts suggests that larger cells may be useful in the anoxic environments of muddy lake bottoms, where salamanders such as *Necturus* live; larger cells permit lower basal metabolic rates, which aid survival in oxygen-starved habitats, but also require extra DNA to serve a skeletal function in the correspondingly enlarged cell nucleus. All of this is compatible with slow rates of larval development, but less so with rapid rates (such as those required by amphibians breeding in temporary pools) in which replication of large amounts of DNA might become limiting (Nevo and Beiles, 1991). The question remains as to why amphibians in general have more DNA than other vertebrates, and Olmo (1991) has suggested that the transition from amphibians to reptiles must have included the evolution of as yet undiscovered mechanisms for more efficient control of genome size.

1.2 SPECIAL FEATURES OF THE ANURA

1.2.1 General characteristics

Most of the distinctive features of anuran amphibians are concentrated at the rear end of the animals. The postsacral vertebrae are fused to form a single bone, the urostyle or coccyx, which gives anurans their characteristic squatting appearance; the hindlimbs are relatively elongated compared with the fore-limbs, and of course there is no tail. This unique pelvic architecture enables anurans to exploit several modes of locomotion, including crawling, hopping, running and leaping; it also provides the apparatus for a powerful swimming action, in which the hind limbs generate all the thrust while the forelimbs are folded back passively along the body to minimize drag.

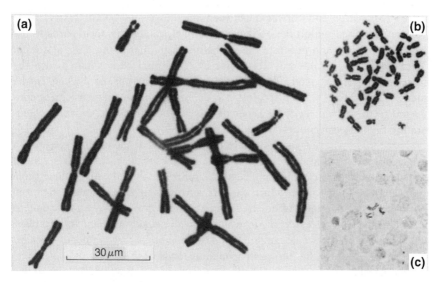

Figure 1.2 Chromosomes of (a) *Triturus cristatus*, (b) humans and (c) *Drosophila*, all to same scale. (Reproduced from Macgregor (1978), *Philosophical Transactions of the Royal Society of London B*, **283**, 309–318, with permission from the Royal Society.)

1.2.2 Sexual dimorphism

It is usually possible to determine the sex of adult anurans by external appear-ance, but the types of secondary sexual characteristics differ considerably between species. Males of some terrestrial frogs have spines or tusks for fight-ing; those of most water-breeding species have nuptial pads (darkened and scleratinized areas of skin for gripping females) on their forelimbs; yet others have conspicuous swollen glands on various parts of their ventral surface, and in a good number of species there are colour differences either on the vocal sac or on other parts (sometimes large areas) of the body that make sex recog-nition easy. A dramatic sexual dimorphism is shown by the hairy frog *Trichobatrachus robustus*, in which males during the breeding season grow long, hair-like protuberances on their flanks and thighs.

1.2.3 Vocal and auditory systems

Unlike the other amphibians, anurans are often noisy animals and thus come equipped for the effective generation and reception of sound. Males are usually louder than females, and the apparatus responsible for creating sound is therefore generally better developed in the former sex. Thus although both male and female anurans have a functional laryngeal apparatus, only males have vocal sacs capable of amplifying the sounds produced by it. The passage of air backwards and forwards between the lungs and buccal cavity, completely contained within the body of the animal, causes the vocal cords in the larynx to vibrate and generate the pulsation of air that is the basis of sound. Vocal sacs serve to intensify this noise by resonation and radiation into the surrounding air or water. They may be no more than crude pockets in the mouth lining (as in some discoglossids), or more sophisticated structures connected to the buccal cavity by a pair of valves (vocal slits) and capable of generating high intensity sound waves.

Three main types of vocal sac have been recognized: the median subgular, a single distension ventral to the throat; the paired subgular, in which a ventral distension is partly or completely divided; and paired laterals. All of these three types may be 'internal', in which case the surrounding skin is unmodified, the noise generated is usually of low frequency, and the area involved swells only to a limited degree; or 'external', in which case the opposite characteristics normally apply. Internal vocal sacs are common in anurans that frequently vocalize underwater, such as the European spadefoot toads of the genus *Pelobates*; terrestrial species and many which call from around or above the water surface capitalize on the greater range achievable with the external sac, which is often so large as to more than double the size of the animal's head when fully inflated.

Although anurans do not possess protruding external ears, most have prominent tympanic membranes that receive sound signals and are obvious to the casual observer as circular patches of skin immediately posterior to each eye. They are especially conspicuous in many ranids, such as the North American bullfrog *Rana catesbiana*. Anuran ears are unusual in possessing two separate receptor systems: one for processing high frequency calls of the type commonly used by males during the breeding season, and another (which anurans share with urodeles) for lower frequency sounds of less than 1 kHz. Only crickets are known to have a similar arrangement.

Anurans were the first animals to evolve vocalization and in many parts of the world their music pervades the night. These amphibians, together with select invertebrate companions, give our world its nocturnal soliloquies. Anuran orchestras are complex affairs: most of the noise is made up of advertisement calls uttered by males either to attract females or to intimidate other males, or both; but there are also release calls, spluttered out when a male or unripe female is grasped by an over-eager suitor, distress calls produced by some species during pursuit or attack by predators, and (much more rarely) reciprocation calls from receptive females in response to male advertisements. The loudest calls are generally produced by advertising males of species which frequent open habitats where noise can travel great distances. The booming *Rana catesbiana* of North America and the strident *Bufo calamita* of Europe can be heard by humans over distances of 1–2 km, while male Australian and North American tree frogs *Litoria ewingi* and *Hyla cinerea* elicit responses from females at least 100 m away (Loftus-Hills and Littlejohn, 1971; Gerhardt, 1975). Anurans breeding in or near flowing water, such as some hylid frogs, call at high frequencies (sometimes greater than 4 kHz) to overcome the acoustic competition of rushing water; conversely, fossorial species which make subterranean calls often utilize lower frequencies (less than 1 kHz) because these transmit better through soil (Tyler, 1976).

The range of sounds that emanate from male anurans is extraordinarily diverse, with whistling, croaking, grunting, warbling, roaring, laughing and popping to name but a few. All this has given rise to a plethora of charmingly descriptive names – bullfrogs, spring peepers, sheep frogs, barking tree frogs and so on. It is also very convenient for identification purposes; just as with birdsong, experienced herpetologists become skilled at recognizing the calls of particular species even among complex choruses, and thus can establish presence without visual contact. More formally, amphibian calls can be analysed structurally by sonographs and used for a range of studies including evolution and hybridization.

Distress calls are uttered by both sexes of many anuran species and can be surprisingly loud and effective. They are often made with the mouth wide open and are more like screams than the familiar croaks and gurgles of the mating season. This kind of vocalization is particularly well developed in ranids and seems to be more common in tropical than in temperate species (Hodl and Gollmann, 1986). Reciprocation calls have been identified only rarely but have been heard from females of the European midwife toad *Alytes obstetricans*, a species in which males advertise from concealed refugia often far away from ponds (Heinzmann, 1970).

1.2.4 Eggs, larvae and metamorphosis

All amphibian eggs are large relative to those of mammals (anuran ova are often several millimetres in diameter) and are contained by a vitelline membrane which is ruptured at hatch by the secretion, from the embryo, of proteolytic hatching enzymes. Anuran ova are produced singly, or in clumps, floating films or long strings, depending upon species, and are always embedded in gelatinous mucoproteins and mucopolysaccharides synthesized in the oviduct. This jelly varies enormously in complexity between species, with anything from one to several distinct layers, and its presence is often crucial for both fertilization and hatching. The ova are frequently pigmented, especially those of species breeding in cold water where the presence of melanin aids heat acquisition and can keep the developing eggs 0.5–2°C warmer than their surroundings. Anuran eggs are highly susceptible to desiccation, though some species minimize this by generating foam nests which develop a crusty, relatively impermeable exterior. The laying of eggs in floating films (as in the bullfrog *Rana catesbiana*) or in long strings (as in toads of the genus *Bufo*) are adaptations to maximize surface area and thus ensure adequate oxygenation.

Anuran larvae are the classic tadpoles, morphologically much more different from their parents than those of the other amphibian orders. Most free-living tadpoles obtain their nutrients primarily by filter-feeding, and much of their food is vegetable in origin. Tadpole intestines are relatively much longer than those of adult anurans in order to cope with digestion of this more refractile material. Various classifications of anuran larvae have been devised, but from the ecological viewpoint only a few types need be distinguished (Duellman and Trueb, 1986): those typical of lentic waters, with minor modifications according to whether they feed primarily at the surface film, in dense vegetation, on the pond bottom or in mid-water; those frequenting lotic waters, usually with more depressed bodies, longer tails and shallower fins than lentic types; and specialized forms with variable morphology, such as those inhabiting bromeliads or which are able to make short journeys across wet ground. Some anurans such as spadefoot toads (*Scaphiopus* spp.) produce tadpoles that can develop as alternative morphs, the larger ones becoming facultative carnivores (even cannibals) and devouring smaller larvae in the same pond. Tadpoles respire by exchanging gases through gills and skin, and sometimes by gulping air from the pond surface. The latter ability varies according to the timing of lung development, which in some cases (such as *Bufo* toads) does not occur until near metamorphosis. Just like their parents, anuran larvae exhibit a range of colour schemes from cryptic to aposematic which are presumably related to mechanisms of defence against predators (Altig and Channing, 1993).

Anuran larval development has been formally compartmentalized into discrete stages by several workers, and that most commonly used (Gosner, 1960) is shown in Figure 1.3.

Metamorphosis is under endocrine control in which the thyroid gland is especially important. The changes concomitant with this dramatic event ramify throughout the animal's body and are often accomplished within a remarkably short space of time. Just occasionally the process goes awry, and tadpoles fail to metamorphose. In the case of European green frogs (*Rana esculenta*), such larvae grow to giant sizes (120 mm, or more than twice the normal maximum) and may survive for several years in this condition.

1.3 SPECIAL FEATURES OF THE URODELA

1.3.1 General characteristics

Newts and salamanders are universally elongate, tailed amphibians which resemble ancestral forms more than do the other two orders. Most have four

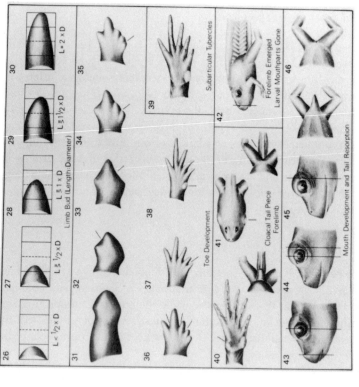

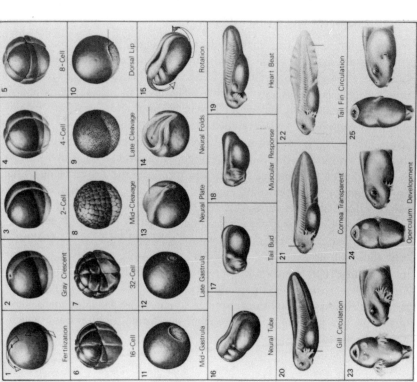

Figure 1.3 Stages of anuran embryonic and larval development. (Reproduced, with permission, from Gosner (1960), *Herpetologica* 16, 183–190.)

well developed limbs, but in some aquatic genera (such as *Amphiuma*) limbs may be vestigial or altogether lacking. Newts are not distinct from other urodeles by any universal criterion, but the term is somewhat arbitrarily applied to a small number of genera (especially *Notophthalmus* and *Triturus*) in which adults repair to water for breeding purposes. Urodeles are generally slow-moving and cumbersome on land, but most swim well and as with anurans there is a complete range of specializations from fully aquatic to fully terrestrial species. Surprisingly, one of the most successful urodele groups in terms of species diversity and abundance, the plethodontids, has become lungless. Although not noted for vocalization, some species of urodeles emit a plaintive squeak when roughly handled.

1.3.2 Sexual dimorphism

Urodeles can be difficult to sex, especially outside the breeding season. During the reproductive period, the cloacal region in males of many species becomes much more swollen than that of females, while the males of European newts (*Triturus*) often also develop large crests and distinctive colorations. All of these features regress when breeding ceases but traces of them usually remain detectable by careful inspection.

1.3.3 Eggs, larvae and metamorphosis

As with anurans, urodeles may produce eggs singly, in clumps or in strings according to species. Those breeding in water often attach them to rocks (sometimes via gelatinous stalks) or fold them in the leaves of aquatic vegetation. Urodele larvae usually resemble miniature versions of adults and are thus more immediately recognizable than those of anurans. Those living in water normally have well-developed external gills and forelimbs from soon after hatching; gills and tail-fins are more pronounced in lentic species than lotic ones, presumably because the latter inhabit well-oxygenated environments and need to minimize surface area to reduce the risk of being washed downstream by strong currents. Another pair of protuberances, the so-called balancers, are also present on the heads of many aquatic urodele tadpoles. Urodele development has been formally described by Harrison (1969) and is shown in Figure 1.4. Respiration is via the gills and skin. Unlike anuran larvae, those of urodeles are exclusively carnivorous: many prey on small invertebrates but some take larger items, including other tadpoles. *Ambystoma tigrinum*, using temporary ponds, has a cannibalistic larval morph which, analogous to the similar situation in spadefoot toads, develops distinctive physical differences from the normal siblings upon which it preys.

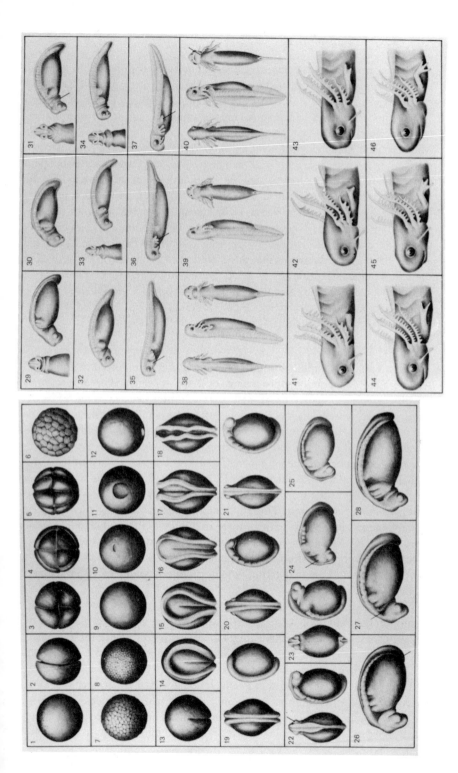

Figure 1.4 Stages of urodele embryonic and larval development. (Reproduced, with permission, from Harrison (1969), *Organisation and Development of the Embryo*, Yale University Press, Connecticut, USA.)

Metamorphosis in urodeles is a less dramatic event with regard to morphological changes than it is in anurans, despite the fact that the larval phase often takes longer in urodeles. Of particular interest is the phenomenon of neoteny or paedomorphosis, in which some aquatic animals retain larval morphology (often including gills) but grow to full adult size and reproduce normally. Many aquatic salamanders (including *Amphiuma*, some *Ambystoma* and the giant *Andrias* species) are obligate neotenates and never metamorphose into terrestrial forms in nature; others are facultative neotenates, including the well-known axolotl *Ambystoma mexicanum*, which can be induced to metamorphose by changing the physical environment or by injection with thyroid-stimulating hormones. Neoteny is sometimes an adaptation to situations in which inhospitable terrestrial environments, especially deserts, surround the aquatic habitat; in other cases the causes are less clear but may be linked (as in the case of the European alpine newt, *Triturus alpestris*) to development in unusually cold ponds or simply a strategy for optimal exploitation of the aquatic habitat when this is particularly favourable.

1.4 BIOGEOGRAPHY

The distribution of amphibians in the world today is summarized in Table 1.2. Numbers are deliberately approximate, reflecting the inevitable uncertainties of some species designations as well as the arbitrariness, in some cases, of intercontinental boundaries. The global pattern has some interesting and curious features: caecilians are perhaps the most straightforward, with their five families widely if patchily distributed across tropical regions, but the nine families of urodeles are markedly concentrated in the northern hemisphere and virtually or completely absent from Africa and Australia. Anurans (20 families) are most diverse in the tropics, and quite dramatically so in the rainforests of South and Central America. This continent has almost twice as many species of amphibian as any other, with all three orders well represented, and not far short of half of all known species in the entire world; at the other extreme, Australia has only frogs. Europe has about a quarter as many species as North America, despite its similar latitude, and the latter continent is particularly rich in urodeles.

Some particular amphibian families are especially successful at present. Among the Gymnophiona, the family Caecilidae contains more than 60% of all caecilians and is represented on all three continents where these animals occur. In the case of urodeles, the Plethodontidae also includes more than 60% of all known species but in this case has only a toehold in the Old World, with almost all members of this family residing in the Americas. The most widespread urodele family is the Salamandridae, which includes the newts, and although it has only about 50 species these are scattered throughout North America, Europe, Asia and Africa. Three families of anurans have more than 600 species, notably the Leptodactylidae (confined to the Americas, mainly South and Central), the Ranidae or 'true' frogs and the Hylidae, which are mostly tree frogs. Both of the latter groups are widely distributed across several continents, as are two other very successful families, the Bufonidae or 'true' toads (with more than 300 species) and the Microhylidae or narrow-mouthed frogs (with more than 250 species).

Table 1.2 Global amphibian distribution

Continent	Approximate number of species			
	Gymnophiona	Urodela	Anura	Total
North America	0	150	80	230
S. and C. America	70	130	1500	1700
Europe	0	25	25	50
Africa	30	<5	880	>910
Asia	40	60	650	750
Australia	0	0	250	250

— 2

Why study amphibians?

2.1 INTRODUCTION

Interest in amphibians seems to relate at least partly to the species diversity of the country in question, but is rarely comparable with that in other vertebrate groups. In Britain, for example, the Royal Society for the Protection of Birds had a membership in excess of 800 000 in 1993; by contrast, the British Herpetological Society managed just over 1000. Numbers, however, rarely tell the whole story and herpetologists often make up for their scarcity by uncanny dedication to their subject. One extreme example of such devotion was the eminent German herpetologist Robert Mertens who was bitten, at the age of 81, by a venomous pet snake. His last act was to keep, for as long as possible, a meticulous record of his developing and ultimately fatal symptoms. More significantly, there is no doubt that the study of amphibians has illuminated many areas of the biological sciences, including ecology but much else besides. These animals deserve, and are beginning to receive, rather wider interest and attention in both amateur and professional circles than has hitherto come their way.

2.2 THE BROADER SIGNIFICANCE OF AMPHIBIANS IN BIOLOGY

2.2.1 Development and differentiation

It was recognized well over a century ago that the production of many and large unshelled eggs, and their subsequent embryogenesis outside the mother's body, made amphibians attractive subjects for the investigation of vertebrate development and differentiation. Studies on this topic are legion, and include the dramatic discovery that nuclei carefully removed from body cells of adult frogs and injected into enucleated eggs can sustain complete and proper development of those eggs (Gurdon, 1974). In this way it was demonstrated that every cell in an adult vertebrate keeps throughout life all the information needed to programme the development of the whole animal; no genetic information is lost during the complex process of cell differentiation. Another result of this work was that frogs and newts were among the first animals to be cloned: nuclei taken from somatic cells of one individual and put back into enucleated eggs generate a series (as many as you like) of genetically identical progeny. Many of these experiments have been carried out with the frog *Xenopus laevis*, largely because it is fully aquatic and easy to keep and breed under laboratory conditions; indeed, *Xenopus* is in many ways the amphibian equivalent of the laboratory rat. Frog eggs are provided with vast stores of the materials they need to carry them through early development (that is why they are so large), and a number of specialized biochemical pathways occur in the growing oocyte which generate these materials. Mammalian eggs, succoured by nutrients from the mother, do without much of this paraphernalia; so although *Xenopus* will no doubt continue to provide valuable insights into one of biology's greatest outstanding problems, the amphibian role model for development will never be fully applicable to other animal groups. Amphibians, especially urodeles, also retain into adulthood much greater powers of tissue regeneration than do mammals and for this reason have long been studied on this subject too. For the ecologist this is a mixed blessing since rapid regrowth of digits (or even whole limbs) makes the long-term marking of individuals problematic.

2.2.2 Cytogenetics

Although in many ways unsuited for classical genetic studies, amphibians have proved invaluable in cytogenetics (the study of chromosome structure) because their cells have large nuclei with spectacular chromosomes that are relatively easy to manipulate (e.g. Macgregor and Varley, 1983). Exquisite examples are the 'lampbrush' chromosomes of oocytes, in which stretches of chromatin are unravelled and readily visible under the light microscope, and the nucleolar ribosomal genes in which gene activity (the transcription of DNA into RNA) was first visualized by Oscar Miller in his famous 'Christmas tree' pictures taken under the electron microscope more than 25 years ago. Increasingly sophisticated molecular biological techniques are being applied to these cooperative amphibian chromosomes to demonstrate the precise location of particular genes, and the times during development that they are switched on and off. There is no doubt that this experimental system will yield more exciting news in the years ahead.

2.2.3 Application-led amphibian research

Perhaps the most widely applied use of amphibians to date was of *Xenopus* frogs in pregnancy tests during the 1950s and 1960s. These tests were based on the response of female *Xenopus* to injection of minute amounts of human chorionic gonadotrophin (excreted in the urine of pregnant women during the first few months of gestation) by producing egg masses less than 24 hours later. Of course colour reactions to test urine are simpler still, and can be carried out at home without tanks full of frogs, so this particular use of amphibians proved fairly transient but was extremely valuable in its day.

Now it looks as if a new generation of frogs might provide a very different service to humankind. The looming crisis of multiple antibiotic resistance in pathogenic bacteria seems unlikely to be countered by tinkering with drugs such as the penicillins that have proved successful in the past, and researchers are increasingly looking towards totally novel sources of bactericidal agents. Frog skins may turn out to be such a source. Recent studies suggest that they have evolved a range of toxic chemicals for just this purpose, and that at least some of the compounds they produce may help us keep one step ahead in the fight against bacterial and perhaps even viral infections that can still return to devastate human populations.

2.3 AMPHIBIANS IN ECOLOGY

2.3.1 Introduction

Ecology is arguably the most challenging and difficult of all the natural sciences, concerned as it is with vast numbers of interactions between highly complex components. Although underpinned by the major biological theory of our time, neodarwinism, it has no all-embracing theory of its own and (for example) cannot yet offer any meaningful explanation of the total species diversity on earth. There is of course a substantial body of partial theories in ecology, covering important areas such as food webs, competition and predation, and it is towards the testing of these that much ecological research is directed. Amphibians are in many cases particularly well suited as subjects for such investigations, for a number of related reasons.

2.3.2 Practical considerations

(a) Ease of capture

A simple but important point is that many amphibians are big enough to see easily but are also slow-moving and live in readily accessible places. Catching urodeles, for example, is, if not as easy as falling off a log, often just a question of looking under one. Sophisticated equipment is rarely needed and a range of simple methods have been developed for making contact with amphibians on land and in the water. Systematic searching by night with a powerful torch is a popular method for finding amphibians foraging on land, especially in open habitats or on roads and tracks (e.g. Denton and Beebee, 1992). Apart from inspecting possible refugia such as logs, stones and burrows, amphibians during their terrestrial phase of life can also often be caught in buckets (equipped with drain holes) set into the ground to act as large pitfall traps (e.g. Strijbosch, 1980a). As with most other methods, sample bias must be taken into account since species vary in their vulnerability to any particular type of capture. An extension of the pitfall approach has been widely employed for catching amphibians on migration into and out of their breeding ponds. In this case, the pond is surrounded by a low plastic drift fence dug some way into the ground so amphibians cannot cross the barrier (Figure 2.1); pitfall buckets are set at intervals along it, and on both sides, to catch animals that meet the fence and then blunder in while trying to enter or leave the pool (e.g. Gittins, 1983). The amphibians are collected at intervals and subsequently released on the other side of the fence to that on which they were caught. Studies of this kind have been popular for estimating sizes of amphibian populations, peak migration times and survival rates during occupancy of the breeding pond. However, great care is needed to make the fence amphibian-proof (newts are excellent climbers) and to watch out for predation of amphibians caught in the pitfall traps. Despite these reservations, this remains a very useful approach and has recently been extended to terrestrial habitats, for example to enclose arbitrary areas of land during winter and see how many amphibians are hibernating there when they emerge and get caught in spring.

Figure 2.1 Pond with drift fence system. (Photo: F. Slater.)

In the water, commonly used methods for adults and larvae include inspection by torchlight after dark (when many species can easily be seen and caught near the pond margins), trawl or hand-netting, and underwater trapping.

Bottle traps (Figure 2.2), which need no bait, have been used very successfully to study newt populations (Griffiths, 1985). Just why newts enter these traps, which are designed rather like miniature lobster pots, is unclear. Although they tend to bias in favour of catching males, traps are effective for all the European newt species and are usually set overnight (with an air bell to prevent suffocation) for inspection the following morning.

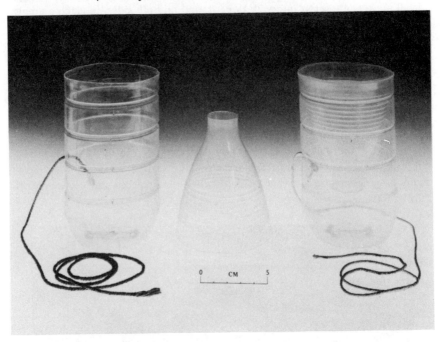

Figure 2.2 Bottle trap for catching urodeles. Components left and centre; assembled trap on right. (Photo: R. Griffiths. Reproduced, with permission, from Griffiths (1985), *Herpetological Journal* 1, 5–10.)

For those amphibians that spawn in ponds, assessing reproductive output is also possible to an extent not realizable with many other taxa. Eggs can be counted in the clumps or strings deposited by the females of many species, and in turn the number of these clumps or strings, which are often laid in conspicuous places and easy to find, can give a measure of the reproductively active female population size (e.g. Cooke, 1975a). Much the same applies to the foam nests produced in or near water by some tropical frogs, such as the *Leptodactylus* species of the Caribbean islands. Even amphibians which lay their eggs singly and wrapped in plant leaves, as many newts do, are accessible to studies of reproductive output by placing artificial oviposition sites (such as flexible plastic leaves attached to a central pole) in a pond and recovering them later for inspection.

A final point on this topic is that virtually all amphibians are user-friendly to human handlers. Most species have no offensive weapons worth talking about, and though a few can bite, serious problems do not arise. The skin toxins for which amphibians are renowned are rarely produced in response to gentle handing, and the worst that researchers will usually experience is a comical defence posture or a shrill alarm call uttered with the mouth wide open. Probably more of a risk are the animals sometimes found together with amphibians, or in similar refugia. In many countries prospecting burrows for toads is at least as likely to produce scorpions or venomous snakes.

(b) Marking individuals

Once found, adult amphibians can be identified on subsequent recapture using a greater variety of methods than can be applied to most other animal groups. For short-term studies, such as breeding behaviour over a few days or weeks, frogs and toads can be fitted with coloured plastic waist-bands, and newts with rings; but these are often lost and can also put the animals at risk by trapping them in vegetation or making them more obvious to predators, although plastic knee-tags have been used successfully to mark frogs (Elmberg, 1989). Historically, the most widely used technique for anurans has involved clipping off the terminal digits, according to a predetermined code, from one or more toes. This does not work well for urodeles because their toes usually regenerate quickly, but for anurans the mutilation is permanent. The method is cheap, quick and easy with the added advantage that bone taken from the digit can be sectioned and stained (at least in temperate species which undergo hibernation) to reveal growth rings, and hence the age of the individual (e.g. Hemelaar, 1981), or used for DNA extraction and genetic analysis. However, there are increasing worries about the pain inflicted and also about subsequent survival. Studies on the effects of toe-clipping on *Bufo woodhousei fowleri* in the United States (Clarke, 1972) and on *Bufo calamita* in France (Golay and Durrer, 1994) both indicated increased mortality and other ancillary problems following the treatment. High-pressure injection of non-toxic dyes ('pan-jetting') to form distinctive subcutaneous marks has been used successfully with toads in Britain, but the method requires some expertise or the animal can be damaged, and in any case the marks fade within a year or two. Cryogenic scarring, widely used with fish, also works with amphibians but again the mark fades quickly in most cases due to the high regenerative power of the naked skin. Increasingly popular in recent years has been the use of natural markings, especially spot patterns on the belly or throat (e.g. Hagstrom, 1973). These tend to be unique to the individual (Figure 2.3) and can be photographed or even photocopied. For small populations (100 animals or less), using colour patterns is probably now the method of choice and fieldworkers often become skilled at recognizing a good many individuals by memorizing these morphological characteristics. Larger populations are much more difficult, needing equipment to record individual patterns and lots of time to scan photographic collections to identify individuals. Computerization of the recognition process may provide a solution but is unlikely to be cheap or simple.

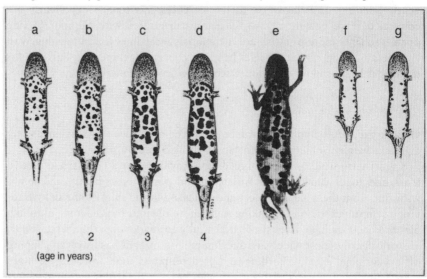

Figure 2.3 Belly patterns of individual *Triturus cristatus* showing changes during growth: (a)–(e) one individual; (f), (g) two weakly-spotted juveniles. (Reproduced, with permission, from Arntzen and Teunis (1993), *Herpetological Journal* 3, 99–110.)

Amphibians can also be monitored using radiotelemetric techniques. Species of average size and above (those weighing 15 g or more) can be rigged with packs that amount to less than 20%, and usually less than 10%, of their body weight and which do not seem to have measurable effects on behaviour. Amphibians are difficult to harness, but transmitters can be implanted in the abdominal cavity under anaesthesia (e.g. Sinsch, 1988) quickly and with minimal risk of mortality. In this way 15–20 g amphibians will carry transmitters with batteries that signal for at least 100 days, and larger animals with ones that go on for much longer. The unit is removed surgically before the expiry date and, provided that operations are not performed near hibernation time, the animals seem to recover well and live normal lives thereafter. A recent development particularly suitable for small amphibians is the use of transponders, which are small metal tags weighing only 100 mg that can be implanted subcutaneously (by injection) and which return an individual-specific signal by reflection from a transmitter. They need no battery and thus have an infinite lifetime, and can be employed with animals as small as 2 g (e.g. Fasola,

Barbieri and Canova, 1993). The main limitation of transponders is their range; whereas transmitters in a 20 g ground-living frog can be detected up to 25 m away, the transponder needs to be within 20 cm or so. This is obviously a serious restriction for many fieldwork studies.

Tadpoles can also be marked. Non-toxic dyes have been used successfully with anurans, but more commonly the tail-fins are clipped with sharp scissors. The mutilation needs to be small enough not to interfere with swimming ability, but sufficiently distinctive so as not to be confused with accidental damage of the kind tadpoles often suffer from predator attack. W-shaped notches can be created quite quickly and serve the purpose well (e.g. Banks and Beebee, 1988). Most problematic are juvenile, immediately post-metamorphic amphibians. These are too small in most species for any of the above methods; toe-clipping of *Bufo bufo* metamorphs has been attempted with some success in Britain (Reading, Loman and Madsen, 1991); small wire transponders are feasible in some species; and cohorts of metamorphs are markable by absorption of suitable radioisotopes, but safety considerations are increasingly likely to limit the application of this type of method.

(c) Advantages of limited mobility

Many amphibians have limited mobility and therefore rather small home ranges. Frogs and toads often stay within remarkably small areas for most of their lives; thus the range of *Rana clamitans* in North America is typically less than 200 m^2, and the tiny *Dendrobates pumilo* is usually content with no more than 20 m^2. Salamander ranges also tend to be small; Plethodontids in North America may use 10–80 m^2, and the European *Salamandra salamandra* only 10–15 m^2 (all as reviewed in Duellman and Trueb, 1986). Amphibians which reside along stream banks, such as the salamander *Desmognathus fuscus* and the tree frog *Hyla cadaverina*, tend to have linear ranges along the stream valleys but these too are often surprisingly small (frequently just 1–10 m). Some species do go further afield, but in general it is relatively easy to keep track of individual amphibians for most of the time. Populations often reside within relatively discrete habitat patches and it is commonly possible to make regular contact with a high proportion of the individuals present. This rather sedentary behaviour pattern of amphibians together with the relative ease of finding them has facilitated studies aimed at identifying key microhabitat components, and also makes amphibians useful models for the study of homing behaviour and orientation. Many species have surprisingly well-developed abilities to return home after displacement (e.g. Sinsch, 1991) and it is much less difficult to follow a newt than to keep track of a pigeon.

A further advantage for researchers is the seasonal migration of adults of many species, especially in temperate countries, to specific breeding ponds at fairly predictable times of year. This assembly behaviour provides an ideal opportunity to observe and manipulate the animals for any of a variety of subsequent investigations. The size of these populations, and hence congregations, varies substantially between species and between different sites for the same species, so with a little groundwork a population size can be chosen depending on the type of question to be addressed. Common frog *Rana temporaria* populations in Britain, for example, vary in size from a single pair to over 5000 individuals using a particular breeding pond; crested newts *Triturus cristatus* occur in Europe in numbers which vary from less than 10 to over 1000 per pond; and populations in tropical countries can be truly enormous, with a staggering 45 000 tree frogs (*Smilisca baudinii*) estimated in a pond of less than 1000 m² in Mexico in the early part of the present century.

(d) Ease of use in laboratory or other simulated environments

Because many amphibians exhibit such limited mobility, they are often easy to maintain and study under standardized laboratory conditions which simulate their natural environment but nevertheless permit control of various parameters for experimental purposes. The ability to carry out experiments of this kind is of great importance as a corollary to fieldwork studies, and often the only way to test hypotheses arising from field observations. Thus vivaria housing amphibians in the terrestrial phase of life have been employed successfully in study areas as diverse as burrowing and desiccation risks in toads and aggression behaviour in salamanders, while aquaria have proved invaluable for behavioural ecologists investigating, among other things, mating behaviour and sexual selection in newts and kin recognition in tadpoles. Of particular value has been the development of artificial replicated pond systems (Figure 2.4), in which the interactions between amphibian larvae in mixed communities can be studied in great detail (e.g. Wilbur, 1987). These experiments have provided among the best evidence of any kind concerning the roles of competition and predation in semi-natural microcosms.

(e) Fringe benefits

There are other aspects of amphibian biology which can be useful to ecological researchers. One is the fact that diet can be assessed by lavaging, a simple procedure in which a plastic tube attached to a syringe containing water is pushed down the oesophagus into the stomach; the water is then expelled gently to flush out the stomach contents by regurgitation. This is fairly quick, causes no damage to the animals (it works well even on newts weighing less than 5 g) and gives a relatively unbiased sample since the intake is at an early stage of the digestion process (e.g. Griffiths, 1986). Even tadpoles, at least of some anuran species, can be induced to release their intestinal contents for analysis just by immersion in fresh water (Pavignano, 1989). High fecundity, and the ease with which larvae can be sampled in large numbers, is another feature of many amphibian species that makes them attractive to population geneticists wishing to apply molecular techniques such as allozyme analysis and DNA fingerprinting.

Figure 2.4 Replicated pond system for community ecology experiments. (Photo: R. Griffiths.)

2.3.3 The importance of amphibians in ecology

Because amphibians are widely distributed over all the tropical and temperate regions of the earth except for remote oceanic islands, and also penetrate sub-arctic tundras, they usually feature somewhere in most natural communities. Amphibians are ectothermic and most of their diversity is concentrated in tropical and subtropical regions (though this is not true of urodeles, which peak in the humid forests of north temperate latitudes) but, also as a consequence of their ectothermic metabolism, population densities of amphibians are often as high or higher than those of mammals in all parts of the world.

Amphibians have featured in some classic studies of community ecology and thus made substantial contributions to our understanding of the subject. Most of these investigations have looked at the roles of competition, predation and specialization in the structuring of amphibian communities in areas where multiple species are present. Classic examples include work over many decades on the plethodontid salamanders of the eastern United States (Hairston, 1987) and studies of South American anurans in the Ecuador rain forests (Duellman, 1978). These and other investigations have shown that amphibian species partition very effectively in both time and space, and that this partitioning can be the result of competition, predation or specialization according to circumstance. In general, amphibian community patterns comply with the idea that climatic stability and spatial heterogeneity (as in tropical rainforests) permit the evolution and maintenance of the greatest diversity levels; there is, on the other hand, little convincing evidence that interspecific competition is greater where diversity is high. What may often happen is that amphibian numbers are limited by other factors such as predation or stochastic variations in the environment, and not by availability of resources.

A striking feature of amphibians is just how abundant they can be, an observation which has yet to be widely appreciated in studies of energy flow through ecosystems. The most successful species are both common and widely distributed. In Britain, the anurans *Rana temporaria* and *Bufo bufo* are approximately as numerous as humans in the country as a whole, with total population sizes measured in tens of millions even after several decades of amphibian declines. Locally, *Bufo bufo* lives at densities of more than 20 adults per ha, and *B. calamita* at up to at least 40 per ha; these correspond to biomasses of 500–600 g per ha, which is nearly twice that estimated for the most comparable small mammal, the common shrew *Sorex araneus*, in similar habitats. In tropical rainforests of Costa Rica, anurans of *Eleutherodactylus* species average more than 750 per ha, while in West Africa the toad *Nectophrynoides occidentalis* has been recorded at more than 4000 per ha. Urodele population densities can be just as impressive, with around 250 adult newts *Triturus cristatus* per ha (and a corresponding biomass of around 1.75 kg) documented at one study site in central England; in the United States, plethodontid salamanders regularly exceed 2000 per ha and in one part of New Hampshire the slender *P. cinereus* was present at more than 2500 per ha (Burton and Likens, 1975). In this case total biomass including juveniles, at 1.65 kg per ha, was similar to that of the larger *T. cristatus* in England and again substantially exceeded that of birds and mammals in the same forest. So there can be no doubt that in many ecosystems amphibians must be playing pivotal roles in energy budgets, both as predators and as prey.

Another aspect of the amphibia that is central to their interest for ecologists relates to reproductive strategies and sexual selection (e.g. Shine, 1979; Halliday and Verrell, 1986). Mating may occur on land or in the water; breeding seasons may be short ('explosive') with scramble competition of males for females, or protracted with lek-like systems in which female choice may play a more significant role. Some species may adopt either strategy, according to circumstances. Within the urodeles, newts form a particularly interesting group because, although fertilization is internal, the sexes do not embrace and male success depends upon the persuasion of females by complex display procedures. The variety of these mating systems, together with the relative ease of studying their sexual behaviour both in the field and in the laboratory, have brought amphibians to the forefront of research into the evolution of sexual selection.

2.4 AMPHIBIANS AS HARBINGERS OF DOOM

2.4.1 Global declines

Over the past few years amphibians have attracted attention as subjects of study for much less welcome reasons than those described above. The prospect of a Global Amphibian Decline (GAD) was mooted, as a result of casual conversations, at the First World Congress of Herpetology held in Britain in 1989 (Wake, 1991). Following a crisis meeting in the United States in 1990, a Declining Amphibian Task Force (DAPTF) was set up in 1991 under the auspices of the International Union for the Conservation of Nature and Natural Resources (IUCN) to investigate the matter and recommend any necessary action. Among other things, DAPTF publishes a regular newsletter (*Froglog*) keeping interested herpetologists in touch with developments and relevant research into amphibian conservation issues, and has established a network of regional groups in many countries around the world.

The central hypothesis of GAD is that amphibians are especially susceptible to environmental change and should therefore receive particular attention with respect to both study and conservation. Two features of amphibian biology have been widely cited as underpinning this hypothesis: firstly, their typically naked and permeable skins may make them highly vulnerable to chemical pollutants and radiation; and secondly, the life style of many species, requiring both aquatic and terrestrial habitats to be maintained in suitable condition, is more at risk of disruption than that of taxa with inherently simpler needs.

2.4.2 Evidence for global amphibian declines

Indications of GAD come from several widely separated parts of the world, foremost among which are North America, Central America and Australia, but also including South America and Europe. Some examples are indicated in Figure 2.5.

The North American evidence mainly concerns species living in the higher mountainous districts of the western United States. Several anurans that were previously common and widely distributed in that area have declined precipitously over the past few decades, sometimes to the verge of extinction. Thus the mountain yellow-legged frog *Rana muscosa* has disappeared from many parts of the Sierra Nevada since 1960, including sites in the Yosemite, Sequoia and Kings Canyon National Parks. The Yosemite toad *Bufo canorus* has also declined at high elevations in California; the boreal toad *B. boreas* has been lost from many parts of the Rocky Mountains and, for example, disappeared entirely from 11 locations in the West Elk Mountains of Colorado. In Oregon, about 80% of the populations of the Cascades frog *Rana cascadae* disappeared between the mid 1970s and 1990; the western spotted frog *R. pretiosa* became extinct in about one third of its range in the same state over the same period; and the red-legged frog *R. aurora* has shared a similar fate of drastic decline throughout its range from Colorado to southern California. There is evidence that populations of the widespread leopard frog *R. pipiens* are in difficulty in the Rocky Mountains. In Canada, the mink frog *R. septentrionalis* in New Brunswick, the green frog *R. clamitans* in Newfoundland and the striped chorus frog *Pseudacris triseriata* in Quebec have all been causes of recent concern. Perhaps surprisingly for a continent so well endowed with urodeles, virtually all GAD stories from North America are about anurans. However, there is some indication that populations of the tiger salamander (*Ambystoma tigrinum*) in the Rocky Mountains have also fared badly in recent times.

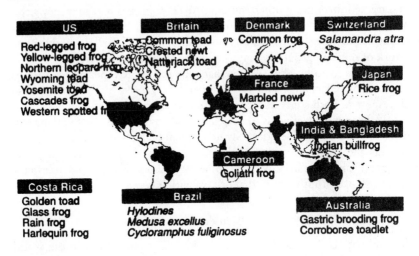

Figure 2.5 Some of the world's declining amphibians. (Reproduced, with permission, from Griffiths and Beebee (1992), *New Scientist* **134** (1827), 25–29.)

The high rainforest of Monteverde in Costa Rica has been the site of another amphibian disaster: the apparent crash in population of the endemic and very beautiful golden toad *Bufo periglenes*. This animal was extraordinarily abundant in the forest in the early 1980s, but seems to have disappeared without trace before the end of that decade.

In Australia, a similar catastrophe seems to have befallen the intriguing stomach-brooding frogs of southern Queensland. These amphibians were only discovered as recently as the 1970s, living along small streams in areas of undisturbed rainforest in the Conondale Range. Their unique mode of reproduction, allowing tadpoles to grow and metamorphose in their stomachs and then regurgitating fully formed froglets, caused great excitement because the mechanism that switched off their digestive processes might also have proved interesting to researchers of human stomach ailments such as ulcers (Tyler and Davies, 1985). However, both the original species (*Rheobatrachus silus*) and the related platypus frog (*R. vitellinus*) can no longer be found anywhere in the wild. Other Australian frogs have also declined or disappeared, including the southern day frog *Taudactylus diurnus* and its congener *T. eungellensis*, the sharp-nosed torrent frog *T. acutirostris* and *Litoria nyakelensis*. The spotted tree frog *L. spenceri* has also decreased in numbers at many of its former haunts in Victoria. These are mainly stream-dwelling species which, interestingly, again live at relatively high altitudes.

Elsewhere, examples have been fewer. In Brazil, eight out of 13 anuran species present in the country's Reserva Atlantica may have become extinct there since 1981 and in Europe there have been many reports of mass mortalities of frogs, mainly the common grass frog *Rana temporaria*, in countries such as Britain and Switzerland.

There are far more accounts of amphibian declines than those cited above, but these examples have formed the main basis for the GAD hypothesis. What many of them have in common is that they occurred in remote places, usually highland areas, where impact from anthropogenic sources is unexpected and not immediately obvious. Another important point is that even at sites where some species seem to be in serious trouble, others in the same pond are obviously not. In the Rocky Mountains, for example, wood frog *Rana sylvatica*, Pacific tree frog *Hyla regilla* and many urodeles are not thought to be in decline. What, then, can be going on?

2.4.3 Causes of global amphibian declines

Since the issue of GAD has only been around for a few years, investigations into causation are still far from complete. A number of interesting facts, however, have come to light and can be classified (rather crudely) in the following way.

(a) Simple causes

One species whose problems may at least be straightforward is the mountain yellow-legged frog *Rana muscosa*. It looks as if the widespread introduction of salmonid fish into mountain lakes of the Sierra Nevada, for sporting purposes, has greatly increased the isolation of *R. muscosa* populations and thus also their chance of extinction by stochastic processes (Bradford, Tabatabai and Graber, 1993). The tadpoles of this frog are highly susceptible to fish predation, and in the absence of fish the adults move along and breed in stream pools. It has been estimated that remaining populations are 10 times more isolated than they were 30 or more years ago, and thus more likely to die out from local outbreaks of disease, increased predation or other factors, without chance of replacement by immigration. In Switzerland, a mass mortality of *R. temporaria* under ice while hibernating at the bottom of ponds during an exceptionally severe winter has been ascribed to anoxia, in other words an essentially natural reason.

(b) Disease

There can be little doubt that the proximal cause of large-scale amphibian deaths in several situations has been infection with fungal, bacterial or viral pathogens. Massive (> 95%) egg mortality in an Oregon population of *Bufo boreas* was observed following infestation of spawn by *Saprolegnia* fungus, and the suggestion was made that this pathogen had been introduced into the mountain habitats along with 'the salmonid fishes that it commonly attacks. However, *Saprolegnia* infection of amphibian spawn is widespread all around the temperate world and fish may not have much to do with this particular problem. Elsewhere, in Colorado, the loss of 11 *B. boreas* populations between 1974 and 1982 has been put down to infection with the bacterium *Aeromonas hydrophila*. This is the best known pathogen of adult amphibians, and causes a distinctive disease with skin lesions known as 'red leg'. There have been numerous well-documented outbreaks of disease, in many amphibian species in Europe as well as North America, attributed to this very widespread microbe. However, recent studies of mass mortalities of the common frog *Rana temporaria* in Britain indicate that, while *Aeromonas* lesions occur in some situations, in others they are rare and other causes of death, possibly due to viral infections, may be equally important. Certainly pox virus particles can be isolated from the skin of animals dying during these disease outbreaks (Cunningham *et al.*, 1993). Many dead and dying anurans in Queensland may also have been the victims of viral infections, particularly *Ranaviruses* of the Bohle Iridovirus (BIV) group, perhaps spread by the release of pet tropical fishes. Finally the recently described post-metamorphic death syndrome (PDS), in which many juvenile frogs die off before or during the spring after their birth, also bears all the hallmarks of an infectious agent. PDS seems to have affected several ranids, especially *R. tarahumarae*, in Arizona and neighbouring areas of the south-western United States.

Diseases are dramatic and their results are easily seen when amphibian populations are large. However, it may be that they usually occur on this scale only under stress-related circumstances that compromise immune systems. This might happen under natural conditions when population densities become high, and thus cause a crash in numbers which will be followed by recoveries and further crashes in an infinite series of chaotic cycles. On the other hand, it might happen as a result of extraneous stresses imposed as a result of environmental deterioration – and this would be more relevant to the GAD hypothesis. If this is true, the environmental factors which promote disease as a secondary consequence need to be identified.

(c) Environmental deterioration

There are several ways in which changes in the quality of their environment, even in remote mountain districts, could affect amphibians. Pesticides constitute perhaps the best-known such hazard; many have been around for more than 40 years, and because some (such as DDT and its breakdown product DDE) are highly persistent, they can eventually crop up in significant concentrations at the very ends of the earth. There is no doubt that amphibians in some places have suffered from pesticide toxicity but no substantial evidence exists to link pesticides with GAD. Indeed, recent reviews of the effects of pesticides on wildlife suggest that, contrary to expectations, amphibians are in general no more sensitive than other groups of organisms that have been tested and in many cases are surprisingly resilient (Hall and Henry, 1992).

Acidification of groundwater by atmospheric pollutants, especially sulphur dioxide dissolving in rain, is another potential cause of GAD which might be expected to impact upon upland hard-rock areas. Again, it is clear that acidification has caused problems for amphibians in Scandinavia, Britain and North America but there is little to connect it with the specific cases associated with GAD. Indeed, studies in mountainous regions of western North America suggest that acidification has not become a serious or widespread problem there and cannot explain the amphibian declines observed in the region (Bradford *et al.*, 1994).

Another problem receiving increasing attention is the damage to stratospheric ozone caused by trace amounts of catalytic chemicals, especially chlorinated fluorocarbons (CFCs), released from industrial processes. The resultant 'ozone holes' centred around the North and South Poles permit greater amounts of solar radiation, especially in the high-energy ultraviolet (UV) range, to reach the earth's surface. One well-publicized consequence has been an increase in human skin cancers, especially in Australia, but the effects on other components of global ecosystems have so far received little attention. However, a recent study of UV-B radiation and Rocky Mountain amphibians provided evidence that an important cause of GAD has been revealed (Blaustein *et al.*, 1994). Filters removing UV-B substantially increased the survival of *Rana cascadae* and *Bufo boreas* eggs under otherwise natural conditions in the breeding lakes, but had no effect on hatch or survival rates of eggs of the sympatric but not declining *Hyla regilla*. Also, levels of the photolyase enzymes which repair DNA following damage from UV irradiation varied in the expected way; *Hyla* eggs had a very efficient repair system, whereas those of *Rana* and *Bufo* did not. At the nearest recording station (Toronto in Canada), at the same latitude as the study site, UV-B levels increased by between 7 and 35% per year after 1989, when recording started. Ozone declined over the same period. The spring months, during which Rocky Mountain amphibians breed, are times of particularly high UV-B irradiation and this is likely to be amplified at high altitudes.

Of the known possible causes of environmental deterioration, damage to the ozone layer and its effects on UV-B irradiation at ground level therefore looks the most likely candidate of relevance to GAD. Even in this case, however, its applicability is likely to be limited. The amphibians that have declined at high altitudes elsewhere in the world are mostly rainforest species unlikely to be exposed in the same way, and some have reproductive cycles in which eggs do not develop outside the mother's body in any case.

2.4.4 Global amphibian declines in perspective

The possibility of GAD has certainly drawn attention to amphibians, but it is important to consider the concept critically and place the fears that have been raised into a more general context. One problem concerns deciding when a particular case should be categorized as part of GAD. This has been somewhat selective, concentrating on declines for which no immediate cause was evident while excluding longer or better studied ones such as that of the natterjack toad *Bufo calamita* in Britain, or the tree frog *Hyla arborea* in much of northern Europe. It will be important to establish that GAD includes everything we know about, even if this demystifies the concept, as it surely will.

A more serious problem is that of distinguishing declines from natural fluctuations in population sizes. One attempt to address this question looked at three urodele populations (*Ambystoma tigrinum*, *A. opacum* and *A. talpoideum*) and one anuran (*Pseudacris ornata*) over a 12-year period in the south-eastern United States (Pechmann *et al.*, 1991). All four species exhibited large (several orders of magnitude) inter-year differences in population sizes, as determined by assembly of adults in, and recruitment of juveniles from, the breeding pond (Figure 2.6). Downward trends over a few years were often followed by complete recoveries, and the four species did not all follow the same trends at the same times. This highlights a major obstacle for conservationists wishing to establish the significance of GAD; there is often no shortcut to long-term studies, and these will be essential if we are to understand what is really happening to animals that can fluctuate naturally in cycles which may have periodicities of several years or even decades (Blaustein, Wake and Sousa, 1994; Pechmann and Wilbur, 1994). This in turn generates a serious dilemma, because to err on the over-cautious side may equally result in delaying a response to a real environmental problem beyond the point at which it is still reversible. Unfortunately it is still not clear whether at least some of the examples cited in support of GAD (such as the golden toad) are really in difficulty or are indeed just experiencing nadirs in wildly fluctuating natural cycles. There are certainly numerous examples of species from other groups coming back, as it were, from the dead: in Britain the scarce emerald damselfly was rediscovered in 1983 after being declared extinct in 1971, the Welsh clearwing moth was rediscovered on Mount Snowdon after more than a century without records, and so on. Whether amphibians ever exhibit this degree of resilience remains to be seen.

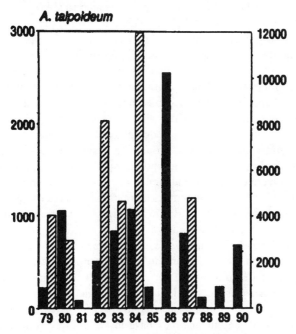

Figure 2.6 Variation in numbers of *Ambystoma talpoideum* between 1979 and 1990 at a pond in South Carolina. Solid bars (left-hand axis) = numbers of breeding females; hatched bars (right-hand axis) = numbers of metamorphosing juveniles. (Reproduced, with permission, from Pechmann *et al.* (1991), *Science* **253**, 892–895.)

What, then, is the overall picture? Of the approximately 4000 species of amphibian known to science, no more than 1% have been cited as contributing to global amphibian declines. As a scientific hypothesis, GAD is essentially untestable because it is logistically impossible to monitor even a reasonable sample of all the species that are out there. We do know that in some places – such as the Caribbean islands, south-eastern United States and lowland tropical forests in Indonesia – there is no evidence of declines in their amphibian populations. We also know that certain species of amphibian, far from declining, are still expanding their ranges. Examples include the marine toad *Bufo marinus*, a native of parts of Central and South America, which has been widely introduced into other tropical countries and usually spread like wildfire as a result; the North American bullfrog *Rana catesbiana*, which in recent decades has expanded beyond its original range in America and also established itself in several European localities; the European marsh frog *Rana ridibunda*, which has appeared in Britain, Switzerland and elsewhere; and the Italian crested newt *Triturus carnifex*, which has been introduced to (and started to spread in) Switzerland. A complete list would be a long one, and probably at least as long as the one of species related to GAD.

Inherent in the notion of global amphibian declines is the idea that amphibians are somehow faring worse than other comparable groups. But are they? At least within the continent of Europe, there is little support for such a concept when the recent fate of amphibians is compared with that of reptiles (Beebee, 1992). Reptiles are also ectothermic, and many are of similar size and live in similar habitats to those occupied by amphibians. Europe has approximately twice as many species of reptile as it does amphibians, but a higher proportion of reptiles have small geographical ranges than is the case with amphibians, and a much smaller proportion of the reptile taxa have really large ranges. Pooling data from multiple European countries, on average 32% of reptile taxa were considered vulnerable or endangered compared with 27% of amphibians. Twice as many reptiles as amphibians were listed on Appendix II (for maximum protection) of the 1979 Berne Convention; eight European reptiles, but no amphibians, were listed on CITES (Convention on International Trade in Endangered Species) Appendices; while four reptiles and five amphibians were listed in the Red Data Book. Overall it looks as if European reptiles are doing at least as badly as the continent's amphibians, but no taxa from either class of vertebrate have become extinct in Europe within historical times. On a global scale, recent analysis of extinction rates since 1600 indicated that amphibians may have lost 0.07% of known species and currently have 2% under threat, in both cases the lowest figures for any vertebrate group (Smith *et al.*, 1993).

Thus, although some amphibian species have certainly declined in recent decades, there is no substantive evidence to indicate that the problems faced by this group are significantly worse than those faced by many others. It is also clear that no single common cause underlies amphibian declines around the world (Griffiths and Beebee, 1992) and that the only useful way to study these declines is to carry out detailed investigations of individual cases (or related groups of cases). None of this should be interpreted as a case for complacency. Many amphibians clearly are declining and amphibian studies can, and frequently do, highlight environmental problems with wide implications for other taxa, including *Homo sapiens*. The apparent disappearance of frogs from seemingly pristine rainforests remains a particularly mysterious problem that we would be most foolish to ignore.

2.4.5 Amphibians as indicators of climate change

Amphibians in temperate regions usually migrate to breeding ponds for spawning in early spring or summer, in annual cycles that are related to maturation of the gonads over the preceding months. Gonad maturation rates are in turn related to ambient temperatures, and the possibility arises that climatic change might eventually affect this process and thus the timing of amphibian reproduction. Evidence from two sites in Britain, visited for breeding by a total of six amphibian species, suggests that such changes may already be underway (Figure 2.7).

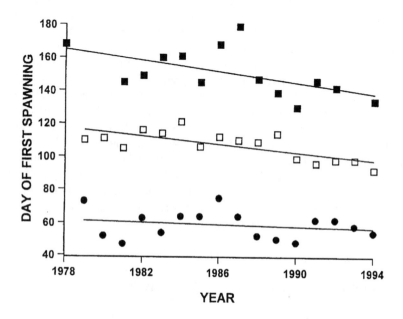

Figure 2.7 Changes in spawning dates of three anurans at a site in southern England between 1978 and 1994. ● = *Rana temporaria*; □ = *Bufo calamita*; ■ = *Rana kl. esculenta*. (Reproduced, with permission, from Beebee (1995), *Nature* 374, 219–220.)

Two out of three anurans spawned progressively earlier over the 17 years 1977–1994, and the first migrants of three urodele species also arrived in the ponds ever earlier over the same period (Beebee, 1995). These spawnings and migrations correlated with temperatures in the immediately previous months, which in turn demonstrated steady increases over the period in question. Amphibian breeding cycles may therefore be useful bioindicators of the consequences of long-term global warming, and surely merit further study in this regard.

— 3

Evolution and phylogeny

3.1 RELEVANCE TO ECOLOGY

Virtually all ecological studies are planned and interpreted in an evolutionary context; questions are asked about the selective advantage of this feature or that behaviour, or how a particular ecosystem has come to take the form that it does. An understanding of the amphibian past, and of the phylogeny of extant groups, is highly relevant to ecological research for at least two reasons. Firstly, it can tell us something about how present distribution patterns came into being; and secondly, it forces consideration of what a species really is. Species are after all the common currency, more or less the 'fundamental particles' of ecology, but the concept is not always straightforward. A good example of a 'species problem' can be seen by comparing lists of European amphibians compiled only 10 years apart (Table 3.1). Arnold, Burton and Ovenden (1978) recognized 43 native species; Corbett (1989) recognized 70. The biological species concept, with its requirement to demonstrate free interbreeding, sometimes cannot be (and elsewhere apparently has not been) applied rigorously to these designations.

Table 3.1 Species list of European amphibians

Arnold, Burton and Ovenden (1978)	Corbett (1989)
Salamandra salamandra	*Salamandra salamandra*
Salamandra atra	*Salamandra atra*
	Salamandra corsica
Salamandrina terdigitata	*Salamandrina terdigitata*
Chioglossa lusitanica	*Chioglossa lusitanica*
Pleurodeles waltl	*Pleurodeles waltl*
Euproctus asper	*Euproctus asper*
Euproctus montanus	*Euproctus montanus*
Euproctus platycephalus	*Euproctus platycephalus*
Hydromantes (=Speleomantes) genei	*Speleomantes genei*
Hydromantes italicus	*Speleomantes italicus*
	Speleomantes flavus
	Speleomantes imperialis
	Speleomantes supramontis
	Speleomantes ambrosii

Arnold, Burton and Ovenden (1978)	Corbett (1989)
Proteus anguinus	*Proteus anguinus*
	Salamandrella keyserlingii
Salamandra (=Mertensiella) luschani	*Mertensiella luschani*
	Mertensiella caucasica
	Neurergus crocatus
	Neurergus strauchii
Triturus marmoratus	*Triturus marmoratus*
Triturus cristatus	*Triturus cristatus*
	Triturus carnifex
	Triturus dobrogicus
	Triturus karelinii
Triturus alpestris	*Triturus alpestris*
Triturus montandoni	*Triturus montandoni*
Triturus vulgaris	*Triturus vulgaris*
Triturus helveticus	*Triturus helveticus*
Triturus boscai	*Triturus boscai*
Triturus italicus	*Triturus italicus*
	Triturus vittatus
Bombina variegata	*Bombina variegata*
Bombina bombina	*Bombina bombina*
Discoglossus pictus	*Discoglossus pictus*
Discoglossus sardus	*Discoglossus sardus*
	Discoglossus galganoi
	Discoglossus jeanneae
	Discoglossus montalentii
Alytes obstetricans	*Alytes obstetricans*
Alytes cisternasii	*Alytes cisternasii*
	Alytes muletensis
Pelobates cultripes	*Pelobates cultripes*
Pelobates fuscus	*Pelobates fuscus*
Pelobates syriacus	*Pelobates syriacus*
Pelodytes punctatus	*Pelodytes punctatus*
	Pelodytes caucasicus
Bufo bufo	*Bufo bufo*
Bufo calamita	*Bufo calamita*
Bufo viridis	*Bufo viridis*
Hyla arborea	*Hyla arborea*
Hyla meridionalis	*Hyla meridionalis*
	Hyla sarda
	Hyla savignyi
Rana temporaria	*Rana temporaria*
Rana arvalis	*Rana arvalis*
Rana dalmatina	*Rana dalmatina*
Rana latastei	*Rana latastei*

Arnold, Burton and Ovenden (1978)	Corbett (1989)
Rana graeca	*Rana graeca*
Rana iberica	*Rana iberica*
Rana ridibunda	*Rana ridibunda*
Rana lessonae	*Rana lessonae*
	Rana camerani
	Rana epeirotica
	Rana holtzi
	Rana macrocnemis
	Rana perezi
	Rana shquiperica
	Rana italica

Some of the increase in numbers of species can be attributed to minor differences in boundaries between the two studies, and some to genuinely novel discoveries (such as *Alytes muletensis*), but much is due to reassessment of taxonomic status. This is particularly marked in the genera *Speleomantes*, *Triturus*, *Discoglossus* and *Rana*, and begs the question of who decides, and on what criteria, how many species there really are.

3.2 METHODS FOR STUDYING EVOLUTION AND GENERATING PHYLOGENIES

3.2.1 Fossil studies

Obtaining a record of past life forms is a potentially powerful way of determining the history of present ones. The problems with the fossil approach to evolution are well known: usually only hard parts are preserved, and the record is always incomplete. Apart from the luck involved in finding useful fossils, there are theoretical reasons for expecting incompleteness. Transient forms living during periods of rapid change will always be much rarer than their longer-lived successors, and many habitats are not conducive to fossil formation. Nevertheless, there is a useful fossil record of the Amphibia.

3.2.2 Studies based on extant species

This approach relies on inferring relatedness and common ancestry based on comparisons of characters measurable in living organisms. These characters take many and increasingly sophisticated forms. Historically, morphological features that are readily quantifiable have been very popular but to these have been added a range of behavioural, genetic, cytogenetic and molecular characters.

(a) Morphology

In amphibians, skeletal measurements and external features, especially secondary sexual characteristics such as tusks and claws in frogs, have been widely used. Difficulties include limited degrees of skeletal variation between closely related taxa, the possibilities of convergence due to selection, and high levels of intraspecific variation in secondary sexual characters.

(b) Behaviour

This has been particularly useful in the case of European newts (*Triturus*) in which the elaborate courtship behaviours can be compared. Unfortunately there are serious problems with quantification of behaviour patterns, which in any case may not be independent of morphology (Halliday and Arano, 1991).

(c) Genetics

The ease with which it is possible to generate viable, fertile hybrids has been taken as a measure of relatedness from which phylogenies can be inferred. This is a convenient experimental approach for anurans, in particular, since fertilization is external and thus can be easily manipulated. Extensive work along these lines has been carried out on, among others, toads of the genus *Bufo* (Blair, 1972).

(d) Cytogenetics

Studies of chromosome number, shape, structure and banding patterns have been carried out in several laboratories. The large chromosomes of amphibians make them particularly accessible to comparative analysis, and differences between species that can be quantified for phylogenetic purposes are readily demonstrable (Green and Sessions, 1991).

(e) Molecular methods

An increasing repertoire of techniques based on comparisons of molecular structures, mostly those of proteins and nucleic acids, is being applied to phylogenetic analysis. There are three main procedures designed to detect differences between proteins that can be related to evolutionary history, and another three based on differences at the DNA level.

Allozyme analysis involves separation of proteins by electrophoresis, followed by selective staining for the products of (usually) more than 20 genetic loci. The method only detects charge differences, and therefore no more than

20–30% of the amino acid substitutions that might have occurred in a particular protein. It is nevertheless a relatively cheap and simple approach, and has been widely developed for phylogenetic as well as population studies with amphibians and many other groups (e.g. Nevo, Beiles and Ben-Schlomo, 1984).

Microcomplement fixation (MCF) involves raising antibodies against a chosen antigenic protein and determining the extent to which they cross-react with the same protein from related species; the assay is based on the ability of **complement** (a critical component of the cellular immune response) to lyse sensitized red blood cells and release haemoglobin, which can then be measured using a spectrophotometer (Champion *et al.*, 1974). The more similar the antigens, the tighter the antibody will bind them and complement, and thus the less complement there is left over to lyse the blood cells. All of this can be quantified as **immunological distance** (ID) and again used to construct phylogenies. The method has been widely used with amphibians and generated much valuable information.

The third protein method, **complete sequencing**, is the most powerful by far and has made major contributions to molecular evolution using universal comparators such as cytochrome *c*. However, it is technically much more demanding than the other two and has not been widely used in the specific context of the Amphibia.

The three main DNA-based methods also vary in their sensitivities, but are potentially more powerful than those based on protein structure because they home in on the genes themselves (e.g. Hillis and Moritz, 1990).

Nucleic acid hybridization measures the stability of homologous DNA sequences from organisms of various degrees of relatedness when they are annealed together *in vitro*. Phylogenetic interpretation is based on the fact that stability decreases as a function of increasing nucleotide divergence (and thus evolutionary distance) between the two DNA strands (e.g. Springer, Davidson and Britten, 1991); it is the oldest of the DNA methods and has been used very successfully in, for example, studies of primate evolution.

Restriction enzyme digestion followed by electrophoretic separation of the products is the basis of **restriction fragment length polymorphism** (RFLP) analysis; polymerase chain reaction (PCR) using random oligonucleotide primers, followed again by electrophoresis of the products, constitutes the **random amplified polymorphic DNA** (RAPD) technique (Williams *et al.*, 1990). Both test, in different ways, sequence variations in a small fraction of the gene or genome under study.

Finally, as with proteins, there is the option of complete sequencing of DNA molecules. This is now substantially easier than sequencing proteins, especially since the advent of automatic sequencing machines which can handle several hundred nucleotides at a single run, and is becoming increasingly popular in many laboratories.

Crucial to DNA sequence analysis, be it partial (e.g. RFLP) or complete, is selection of an appropriate sequence for comparison. This in turn depends on how far distant, in evolutionary terms, the organisms in question are likely to be. For closely related species or subspecies, mitochondrial DNA or the so-called 'internal transcribed spacer' (ITS) region of the nucleolar ribosomal RNA genes are likely to be useful because they change relatively fast; for more distantly related species or genera, the coding regions of proteins or the nucleolar ribosomal genes (especially the 18S ribosomal RNA gene, for which a large comparative database now exists), which change much more slowly, are more appropriate. A particular advantage of the RAPD method is that by judicious use of primers a wide range of change-rates can be assessed, though there can be compensatory difficulties in product identification.

Molecular methods in phylogenetic analysis are appealing because they generate readily quantifiable data, and also because at least in some cases they can be related via a 'molecular clock' to real, geological time. Nevertheless they are not without difficulty, and frequently require untestable assumptions about mutation, neutrality of variation and fixation rates.

3.2.3 Generating phylogenies

Creating a phylogenetic tree is a complex business and there are several different mathematical approaches to the subject (e.g. Beanland and Howe, 1992). Standard methods have been developed for quantifying genetic distances (Nei, 1987); and UPGMA, a form of cluster analysis based on measurements of genetic distance, has proved popular for calculating phylogenies. Ideally, phylogenies are derived using as varied a database as possible (e.g. with morphological and molecular characters) and with several tree inference methods, checked for statistical robustness and compared to see how well the different approaches agree in the final analysis.

3.3 ANCIENT HISTORY OF THE CLASS AMPHIBIA

The first amphibians are thought to have heaved themselves on to dry land during the mid-Devonian period, 360 million or more years ago. It is from this epoch that the earliest quadrupedal fossils date, including the well-studied *Ichthyostega*, which retained several features characteristic of the crossopterygian lobe-finned fishes from which the early amphibians are thought to have evolved. *Ichthyostega* had remnants of opercular (gill-covering) bones and a tail-fin with bony rays, but also fully developed limbs and limb girdles. It seems clear that there must have been numerous antecedents to *Ichthyostega* but fossils of those critical evolutionary stages are few. The carboniferous period (345–280 million years ago) witnessed a dramatic diversification of the early Amphibia (Figure 3.1), coinciding with the spread of dense forests that were later to become fossilized as coal deposits. It was also a period in which the earliest reptiles made their appearance and signalled the beginning of the end of world domination by amphibious vertebrates. Many groups died out during the Permian epoch (280–225 million years ago), but some of the largest amphibians ever known, including the 4 m long *Mastodonsaurus* with its short tail and massive, heavily toothed skull, inhabited the swamps of the Triassic age between 225 and 190 million years ago.

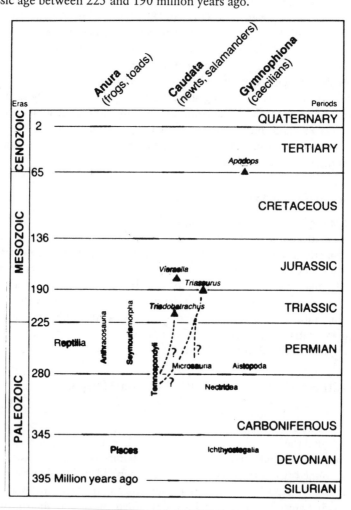

Figure 3.1 Evolution of amphibians. (Reproduced, with permission, from Halliday and Adler (1986), *The Encyclopaedia of Reptiles and Amphibians*, published by Andromeda.)

Almost as perplexing as the origins of the first amphibians are those of the surviving extant groups. The poor fossil record suggests that anurans and urodeles appeared in the mid-late Triassic, but no caecilian remains older than the early Tertiary, 65 million years ago, have yet come to light. It remains contentious as to whether the three living orders derive from a common ancestor (and should thus be grouped together in a single subclass, the Lissamphibia), or whether they had separate origins among their Permian or Triassic forebears. Certainly anurans, urodeles and caecilians share a number of features that seem unlikely to have arisen independently, including elements of tooth, ear and eye structures, fat bodies associated with their gonads, a moist, heavily glandularized skin and a buccal respiration system. On the other hand, some structures are very different in the three orders; this is particularly true of the vertebral column, skull and jaw musculature. More and better fossils may ultimately resolve this issue, but at present the most persuasive evidence seems to favour a 'Lissamphibian' common origin for all the modern species.

Finally, studies of extant species clearly indicate that the closest living relatives of modern amphibians are the dipnoans (lungfishes). Dipnoans, together with crossopterygians, were abundant and diverse in the Devonian period during which amphibians arose. It therefore remains possible, though on the balance of evidence unlikely, that ancient dipnoans were the ancestors of amphibian stock. Further details concerning the evolutionary history of the amphibia are given in Duellman and Trueb (1986) and Halliday and Adler (1986).

3.4 PHYLOGENETIC STUDIES OF URODELES

3.4.1 Introduction

The urodeles are replete with examples of difficult phylogeny, and in Europe the genera *Speleomantes*, *Salamandra* and *Triturus* in particular have generated disagreements that have spanned decades. These creatures exemplify the problems faced by ecologists seeking to put definitive names on the organisms they study.

3.4.2 European fire salamander *Salamandra salamandra*

The fire salamander is one of Europe's largest and most beautifully marked urodeles. An inhabitant (for the most part) of moist deciduous woodland, it ranges in some form or other over virtually all of south and central Europe as well as parts of North Africa and the Middle East. Steward (1969) recognized eight subspecies endemic to Europe, including four confined to the Iberian peninsula; this differentiation was based primarily on spot or stripe patterns (the dorsal surface of all subspecies is black, strikingly embellished with yellow or gold markings), but also other morphological features including overall size and body proportions.

Recently, attempts have been made using more objective criteria to differentiate the various morphs of *Salamandra*. Bosch and Lopez-Bueis (1994) applied quantitative statistics to the dorsal coloration of six populations of animals expected to represent two of the Iberian subspecies: the widespread *S. s. bejarae* and much more localized *S. s. almanzoris*. By this method it was possible to ascribe one population of *bejarae* to the more northerly subspecies *S. s. terrestris*, and another to *almanzoris* – a useful demonstration that comparative morphology, carefully applied, can be just as profitable as the newer molecular techniques. By contrast, Veith (1992) carried out an allozyme-based study in the region in Germany where the two most widespread morphs, *S. s. salamandra* and *S. s. terrestris* (Figure 3.2), overlap and interbreed. He was able to show that, within a broad (140 km) mixing zone, populations of *Salamandra* were panmictic and no barriers to interbreeding (as opposed to migration) between the two subspecies were operating. The current situation reflects the reassociation of forms previously separated during the last Ice Age, and the differences between them are unlikely to be stable in the long-term future.

3.4.3 European newts of the genus *Triturus*

Twelve species of the genus *Triturus* are currently recognized in Europe, and these newts have in recent years become something of a paradigm for phylogeneticists. Studies have been made at two distinct levels of scale; the first, arguably of greatest interest to ecologists, concerns the resolution of subspecies from species, whereas the second attempts to clarify relationships between full species in the genus as a whole.

As with *Salamandra* and many other European amphibians, current newt distributions at least partly reflect migration northwards from refugia around the Mediterranean basin as the ice receded some 10 000 years ago (Figure 3.3). The three most successful recolonists – the smooth newt *Triturus vulgaris*, the alpine newt *T. alpestris* and the crested newt 'superspecies' – all exist in a variety of morphs that have been variably classified as subspecies or full species. In the case of *T. vulgaris* and *T. alpestris*, diversity is particularly high in the Balkans (the presumed main Ice Age retreat of both taxa). Molecular (allozyme) and morphological criteria have been applied to make sense of a complex pattern of variation in *T. vulgaris*, proposing the acceptance of seven subspecies (two of which are not European) that can interbreed freely and none of which merits elevation to species status (Kalezic, 1984; Raxworthy, 1990). Just one of these subspecies, *T. v. vulgaris*, is widespread and accounts for perhaps 90% of the global distribution of smooth newts. A similar clarification, with elimination of some of the 10 subspecies previously proposed, is well under way for *T. alpestris*.

Figure 3.2 Specimens of *Salamandra s. salamandra* (above) and *S. s. terrestris* (below). (Reproduced, with permission, from Steward (1969), *Tailed Amphibians of Europe*, published by David & Charles.)

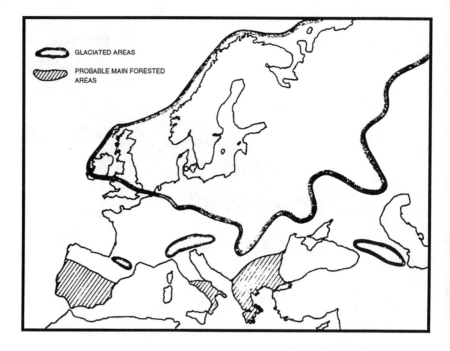

Figure 3.3 Extent of ice cover during the most recent European glaciation. (Reproduced, with permission, from Steward (1969), *Tailed Amphibians of Europe*, published by David & Charles.)

By contrast, a similar exercise with crested newts elevated what were previously considered to be four subspecies up to full species ranking (Bucci-Innocenti, Ragghianti and Mancino, 1983). The change was based on a combination of chromosome and hybridization studies, and was later supported by investigations of mitochondrial DNA RFLPs in border zones where pairs of the four species meet, and which indicated little or no genetic introgression between what are now *Triturus cristatus*, *T. carnifex*, *T. karelini* and *T. dobrogicus* (Wallis and Arntzen, 1989).

At the higher level, there has been considerable progress in resolving the relationships between the 12 full species of *Triturus*. Individual studies have concentrated on various specific aspects such as morphology and behaviour but it is the synthesis of these and other characters (especially cytological and molecular) that has been particularly impressive. This has resulted in a consensus phylogeny in which only three trichotomies remain uncertain (Macgregor, Sessions and Arntzen, 1990; Halliday and Arano, 1991), as outlined in Figure 3.4. Another conclusion of this considerable achievement was that the main radiation of the *Triturus* group probably occurred during the Miocene period, between 25 and 5 million years ago.

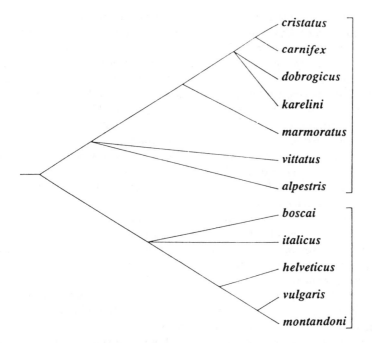

Figure 3.4 Phylogeny of European newts of the genus *Triturus*. (Reproduced, with permission, from Halliday and Arano (1991), *Trends in Ecology and Evolution* 6, 113–117.)

3.5 PHYLOGENETIC STUDIES OF ANURANS

As in the case of urodeles, investigations into the phylogenies of anurans are increasingly important to ecologists in clarifying the taxonomic status of their subjects. Examples are legion. MCF analysis of the Australian stomach-brooding frogs has shown that the two (perhaps now extinct) species of *Rheobatrachus* are not closely related to any other extant genera (Hutchinson and Maxson, 1987), while extensive work on ranids has clarified taxonomies all over the world. However, one of the major achievements of amphibian taxonomy has stemmed from the massive effort put into toads of the genus *Bufo*. Undoubtedly among the most successful of all animal genera, there are currently more than 200 accepted species spread across the entire globe excepting only the Polar regions, Australasia (where *B. marinus* was introduced by humans), Madagascar and Oceanic islands. A range of studies on this genus including fossil history, internal and external anatomy, karyotyping, vocalizations, interspecific (including intercontinental) hybridizations and some biochemical comparisons of blood proteins and skin toxins was assembled by Blair (1972) and the data were used to synthesize a proposed evolutionary history of the 'true toads'. This evidence indicated an origin in South

America, from leptodactylid progenitors, possibly in the Oligocene period 38–25 million years ago, and an early differentiation into narrow- and broad-skulled toads, the former including the most cold-tolerant species and the latter those essentially confined to tropical or subtropical regions. This early diversification was followed by expansion of both groups into North America and across the Bering land-bridge into Asia, Europe and Africa. Some of the radiations of *Bufo* within the Americas were mapped by Blair with confidence (e.g. the *americanus* group), but the situation in the Old World was less clear. The Eurasian species seemed to represent separate lineages, and the narrow-skulled toads of Africa were difficult to place. Evidently the broad-skulled *Bufo* underwent a major diversification in Africa, including the generation of the 20-chromosome toads (notably the *regularis* group) which differ from all the other *Bufo* species that have 22 chromosomes.

A substantial reappraisal of *Bufo* evolution followed the application of the MCF method in an extensive molecular comparison of extant species from all the major continents. This comprehensive study culminated in two major developments (Maxson, 1984). Firstly, the origins of *Bufo* were put back much earlier than previously thought, to some time during the Cretaceous period 140–70 million years ago. At that time South America and Africa were still fused as a single unit, Gondwanaland, and early bufonids could have occupied both land masses before they finally split apart some 95–88 million years ago. This would also account for the dramatic diversification of *Bufo* in Africa, which is in many ways comparable with that in South America. Secondly, the molecular analysis did not support any fundamental differences between broad- and narrow-skulled toads, a distinction which it now seems may have little evolutionary significance.

The *Bufo* studies exemplify the value of molecular methods in systematics and phylogenetics. Although not without their critics, they undoubtedly have powers unparalleled by the alternatives and when carefully applied can revolutionize our understanding of evolutionary histories. The MCF 'molecular' phylogeny of *Bufo* is shown in Figure 3.5. Other molecular methods, especially allozyme analyses, have continued to refine our knowledge of *Bufo* (e.g. Nishioka *et al.*, 1990) and seem likely to maintain this genus as one of the best documented of all animal groups. Another important general point has emerged from molecular studies of phylogeny: morphologically similar amphibians are often much more distantly related than they appear.

3.6 HYBRIDS

3.6.1 Introduction

As a general rule, attempting to mate with a member of a different species amounts to a waste of resources by both partners and strong selection normally operates to prevent it. In the case of sympatric amphibian species a variety of isolating mechanisms have evolved to achieve this end, including differentiation of breeding seasons in space and time and sophisticated recognition behaviours. These fail so rarely that exceptions are memorable; thus a single natural hybrid between the widely sympatric smooth and palmate newts, *Triturus vulgaris* and *T. helveticus*, was thoroughly characterized after its discovery in a Welsh pond (Griffiths, Roberts and Sims, 1987) and extensive interspecific pairing between male frogs *Rana temporaria* and female toads *Bufo bufo* was documented during an unusually cold spring in southern England (Reading, 1984). However, situations exist in nature in which hybridization is much more important and there are some spectacular examples of this among the Amphibia.

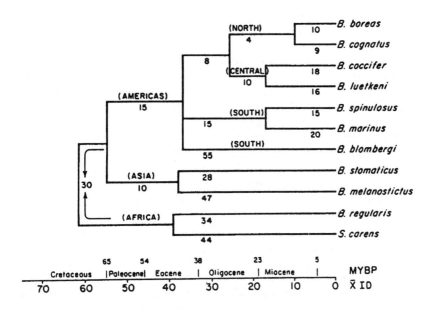

Figure 3.5 Phylogeny of toads genus *Bufo*. The scale indicates Immunological Distances. (Reproduced, with permission, from Maxson (1984), *Molecular Biology and Evolution* **1**, 345–356, published by the University of Chicago Press.)

3.6.2 Hybrid zones

Repopulation of much of Europe and North America by closely related amphibians following the last Ice Age resulted in two types of outcome: either reproductive isolating mechanisms proved adequate to permit extensive sympatry and widely overlapping distributions, or species became demarcated by hybrid zones in which interbreeding still persists and often (but not always) produces progeny of low viability or fertility. Several such zones occur and have been intensively studied; mostly they are rather narrow, but there is considerable variation and some extend over several hundred kilometres. Among

European urodeles, the best known is the protracted hybrid zone stretching across much of France, and up to several hundred kilometres wide, between *Triturus cristatus* and the marbled newt *T. marmoratus*. This is a sympatric hybrid zone (because both parent species intermingle without a cline of intermediate hybrid forms between them) and seems to be moving slowly southwestwards, with the former species extending its range at the expense of the latter (Schoorl and Zuiderwijk, 1981). Hybrids are morphologically distinct, so much so that they were once given specific names, but are sterile.

A rather different situation is exemplified by the yellow and fire-bellied toads *Bombina variegata* and *B. bombina* in eastern Europe, and the American and Dakota toads *Bufo americanus* and *B. hemiophrys* in North America (e.g. Szymura and Barton, 1986; Green, 1983). Both of these pairs meet in long but narrow areas (10–20 km wide) of intense interbreeding, in which allopatric hybrid zones contain a clinal gradation of morphs intermediate between the two parent stocks. Hybrids are fully fertile, though at least in the *Bombina* case the spawn of mixed pairs suffers high mortality rates in the central region of the zone. Allozyme analysis (and also mitochondrial DNA analysis for *Bombina*) has demonstrated genetic introgressions between both pairs which extend over zones perhaps 10 times wider than the 10–20 km cores. Another common feature of these zones is that, unlike that for the newts described above, both occur in regions of habitat transition between ecotypes which favour one or the other parent species, though this transition is much sharper for *Bufo* than for *Bombina*.

Yet another type of hybrid zone occurs between the natterjack and green toads, *Bufo calamita* and *B. viridis*. These animals have ranges which overlap in central and eastern Europe by some 1200 km, including virtually all of Germany and Poland. The two species frequently share similar habitats and breeding seasons, and interbreeding can then occur, although at frequencies much lower than expected by chance. Clearly some mechanism, probably behavioural and based on the very different male advertisement calls, maintains substantial but not complete isolation (Hemmer, 1973). Only one cross (male natterjack × female green toad) produces viable progeny, and these distinctive hybrids are regularly encountered in the wild. In parts of southern Sweden it is feared that the frequency of hybridization between these species is currently too high for both, or perhaps either, to persist (Schlyter, Hoglund and Stromberg, 1991), though whether this situation has only developed recently is not clear.

3.6.3 Hybridogenesis

The significance of hybridization in amphibian biology took a surprising new twist with the discovery by Berger (1973) that the widespread and common European edible frog, *Rana esculenta*, is actually a fertile hybrid of two other green frog species and is now widely referred to as *R. kl. esculenta*. This reflects the novel klepton concept, different from full species but acknowledging the existence of viable, fertile hybrids (Polls Pelaz, 1990). The parent forms are the marsh frog, *Rana ridibunda*, which is primarily an east European animal, and the pool frog, *Rana lessonae*, which occurs in much of central and western Europe but was previously considered to be merely a subspecies of *esculenta*. All these frogs are superficially similar and all three spend most of their lives in or near water, though there are significant morphological and ecological distinctions between them. Pool frogs are the smallest, with relatively short hind limbs and a tendency to frequent small ponds and ditches; they often overwinter on land. By contrast, marsh frogs are substantially larger, live in deep ponds, lakes and canals and usually hibernate in the water. The complex genetics of these frogs has been reviewed by Graf and Polls Pelaz (1989).

There are several remarkable features of the hybrid *R. esculenta*. Firstly, it is extremely successful – so much so that in most areas it greatly outnumbers *lessonae* and pure populations of this parent are now rare. Secondly, most hybrid lines seem to have been produced by crosses between male *lessonae* and female *ridibunda*; the reciprocal mating is much less common, apparently for behavioural reasons. Since males are the heterogametic sex in *Rana*, all the *ridibunda* genomes carried by *esculenta* are therefore female. Thirdly, there is very little genetic recombination (probably less than 5%) between the *ridibunda* and *lessonae* genomes during meiosis of the germ cells in *esculenta*. Sperms and eggs are therefore produced which normally bear one of the parent genomes in unsullied form. This has important consequences, which in turn vary according to the situation in which *esculenta* finds itself.

Edible frogs occur in three main community types. The one most commonly seen in western and central Europe is a situation in which mixed populations of *esculenta* and *lessonae* coexist, usually in small to medium-sized water bodies; this is the so-called **LE system**. Edible frogs in LE systems generate eggs and sperms which contain only *ridibunda* genomes, the unrecombined *lessonae* ones being somehow discarded. Since the *ridibunda* genomes were originally derived from female marsh frogs, it follows that only female *ridibunda* genomes are perpetuated, essentially in clonal fashion, by edible frogs in LE systems. Female edible frogs (genotype *lr*) normally choose to mate with male pool frogs (genotype *ll*); since only *r* is passed on by the edible frog, this generates more edible frogs (*lr*) and no other progeny. Male edible frogs usually have low fertility; when they mate with female pool frogs the outcome is the same as the reciprocal cross except that in this case the progeny are all female (thus generating a female-biased overall sex ratio in edible frog populations), but sometimes they mate with female edible frogs producing *rr* progeny. These, however, have low viability (presumably the clonal *ridibunda* genomes, after multiple generations without recombination, accumulate deleterious mutations) and in any case are always female, and thus incapable of giving rise to self-perpetuating *ridibunda* populations. Nevertheless, the existence of an all-female *ridibunda* stock derived from local edible frogs has in fact been demonstrated by a mitochondrial DNA RFLP analysis at a gravel pit in Switzerland (Hotz, Beerli and Spolsky, 1992).

A different type of situation, in which mixed populations of edible and marsh frogs interbreed in an **RE system**, occurs in central and eastern Europe.

Male edible frogs in RE systems produce gametes containing both types of genomes, normally with l (male) sperm outnumbering r (female) sperm by about 2:1. This leads to a very high male bias in the edible frog component of these mixed communities. Finally a third situation, in which edible frogs exist alone without either parent, has been found, much more rarely, in some parts of Poland, northern Germany and southern Sweden (e.g. Ebendal and Uzzell, 1982). It turns out that a high proportion, frequently more than 80%, of the frogs in these populations is triploid. Sperms and eggs from these animals can be either haploid or diploid, and the latter constitute a vehicle for the perpetuation of *lessonae* genomes in the absence of the parent species.

It seems that other water frog species in southern Europe and eastern Asia may display similarly complex hybridogenic traits, though they have not yet been as thoroughly investigated as the one outlined above. Even in the case of the central European green frogs, many features of the hybridogenic situation remain bizarre, obscure, or both. For example, over large parts of south-eastern Europe there is so large an excess of female over male edible frogs that a special mechanism, involving the production of phenotypic females from genetic males, may be operating in this region (Berger, Uzzell and Hotz, 1988).

3.7 ECOLOGICAL IMPLICATIONS OF PHYLOGENETIC STUDIES

3.7.1 What is a species (and does it matter)?

The increasing realization that boundaries between species can be inconveniently imprecise has generated some challenging questions for ecologists. The problem of species definition is exemplified by comparing the situations of fire salamanders and fire-bellied toads described above. In both cases there are zones in Europe where morphologically distinct forms meet, interbreed freely and produce viable progeny; but the salamanders are considered to be subspecies, the toads full species. The genetic distance between the salamander subspecies was estimated at around 0.121, with a divergence time some 1.6–2.3 million years ago (Veith, 1992); as far as we know, offspring of mixed parentage are fully viable. By contrast, genetic distance between the toads is about 0.487, with a divergence time of some 2.5–6.8 million years ago, and mortality of hybrid embryos can be substantial (Szymura, 1983; Szymura and Barton, 1986). The salamanders, it seems, mix freely and can to all intents and purposes be treated as a single entity, but for the toads it is not so simple. Reproductive success will vary for *Bombina* individuals according to whether they mate homo- or heterospecifically, and the parent species also have different habitat requirements; competitive interactions are quite likely in this border zone. Perhaps the message for ecologists faced with such situations is that high priority should be given to the necessary genetic work before embarking on other types of study that might end up confused by not knowing what is really going on between different morphs. It is also important not to 'create' new species, on flimsy morphological grounds, without checking by molecular and/or genetic means that they really are distinct.

3.7.2 What happens in hybrid zones?

In sympatric hybrid zones, the two overlapping species often occur at high abundance right up to, and within, the zone itself. This happens in the case of the European crested and marbled newts in France outlined above. What, then, prevents one species from overwhelming the other? Outside the hybrid zone, both species have rather similar (and broad) spatial niches, but within the zone there is local allopatry: *Triturus cristatus* dominates the open lowlands and *T. marmoratus* the wooded hills, and only in a few areas is there significant intermixing of species and hybrid formation (Schoorl and Zuiderwijk, 1981). In those ponds where the two species do meet, it seems that *T. cristatus* arrives much earlier in the year than in those ponds where it occurs alone, and also much earlier than *T. marmoratus* in any pond. These early *T. cristatus* are therefore able to breed with minimum risk of hybridization (Zuiderwijk and Bouton, 1987). By contrast, *T. marmoratus* dominates numerically later in the season and presumably benefits at that time in a similar way. It looks as if selection has acted against interbreeding to separate the breeding seasons of the two species temporally, via a special type of competition that has been called 'reproductive self-destruction'. It still is not clear, however, why this situation is relatively stable in space and time. There may be other factors, especially environmental variables such as temperature, which tip the balance in favour of one species or the other on the two sides of the hybrid zone but there is clearly much more to learn about how competition operates in these areas.

3.7.3 What is the significance of hybridogenesis?

The remarkable genetics of the European water frog complex have resulted, among other things, in the *Rana ridibunda* genome spreading (in the edible frog hybrid) into large areas which would otherwise be denied it in its diploid (marsh frog) form. Thus hybridogenesis in LE systems appears to represent a success for *ridibunda* at the expense of *R. lessonae*, upon which it is essentially a sexual parasite (Schmidt, 1993). These hybridogenetic frogs may in effect be cyclical parthenogens, perpetuating the *ridibunda* genome clonally for most of the time but occasionally (via edible frog × edible frog matings, which produce female marsh frogs) allowing sexual recombination before further clonal passaging (Figure 3.6). Among the many questions which arise from this situation, an important one is whether the system has somehow been selected to operate in this cyclical way. In other words, can *ridibunda* genomes recognize some type of internal clock and thus influence, perhaps by the mate-selection behaviour of their host, the frequency of clonal and sexual reproduction? Or is it all completely random, in which case we might expect occasional local extinctions of edible frogs in isolated LE systems when the accumulation of deleterious mutations in the *ridibunda* genome reaches some critical level? Further ecological studies on mate selection may help to answer this question. It is already known that proportions among progeny in mixed green frog populations are often different from those of the adults and vary from year to year, which might imply variation in mate selection (Berger and Berger, 1992); but on the other hand it might reflect differential mortality, since (for example) edible frog tadpoles fare better than pool frog ones in the presence of some predators (Semlitsch, 1993).

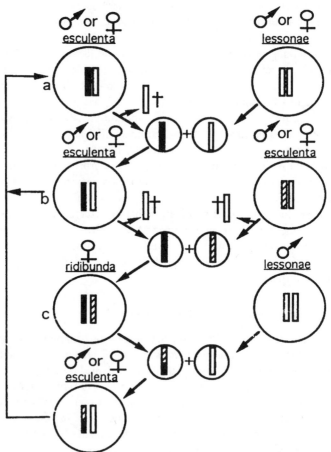

Figure 3.6 Cyclical parthenogenesis in hybridogenetic frogs. Filled and hatched bars denote haploid *ridibunda* genomes; open bars haploid *lessonae* genomes. Large circles are zygotes, small circles are gametes. Recombination is shown in (c) only. (Reproduced, with permission, from Schmidt (1993), *Trends in Ecology and Evolution* 8, 271–273.)

4

Behavioural ecology

4.1 INTRODUCTION

Amphibians are in many ways ideal subjects for behavioural studies; they have limited mobility, are easy to work with in captivity and display a staggering array of interesting behaviour patterns. Three areas in particular stand out as meriting special attention. Firstly, the remarkable migrations that many species undertake to their breeding sites have fascinated naturalists for decades; secondly, the highly varied mating systems of urodeles and anurans have provided useful testing grounds for ideas concerning the operation of sexual selection; and thirdly, the humble anuran tadpoles are being increasingly scrutinized for behavioural evidence of kin recognition and selection.

4.2 MIGRATION

4.2.1 General aspects

Amphibian migration is defined as the act of moving from one spatial unit of habitat (such as winter quarters) to another (such as a breeding pond), and is thus distinct from other types of movement within, for example, a summer home range (Sinsch, 1990a). Most observations on amphibian migrations have been made on temperate species in northern latitudes, but there is no reason to suppose that tropical amphibians are any different in this respect: the Andean toad *Bufo spinulosus*, for example, clearly engages in migrations to breeding sites (Sinsch, 1990b). Nevertheless, it is the spectacular travels of adult amphibians in the north temperate zone, usually in response to seasonal cues, that have particularly caught the imagination. The distances involved can be surprisingly long for such small animals; in general, adult urodeles regularly manage up to 1 km and anurans perhaps 3 km, but a few species go even further and there are reports of *Rana lessonae* in Austria managing an astounding 15 km.

Although most work on migration has concentrated on movement between terrestrial hibernation and aquatic reproduction sites in spring, this is actually only one of at least 10 different types of migration in which amphibians engage (Glandt, 1986). The other nine include travels between summer and winter quarters, aquatic and terrestrial, in various combinations. At least 16 of

19 European amphibians investigated by Glandt exhibited some form of migratory behaviour, but it is the spring journeys that have attracted most attention. Motivation is particularly strong during migrations to breeding sites and this undoubtedly facilitates experimentation.

The speed and accuracy of amphibian migrations pose a number of specific problems. How do the animals 'know' their starting position relative to their goal (the so-called 'map step')? How do they calculate the direction which will take them directly to their goal (the 'compass step')? Are there multiple faculties involved in the accurate orientation so characteristic of amphibian migrations, and do these differ between species? Research over the past 50 years has gone some way towards answering these questions for a few species, but inevitably our understanding of the subject is still far from complete. Standard approaches to the investigation of amphibian migrations include the use of drift-fencing around breeding ponds to assess the timing and direction of immigration and emigration, and deliberate displacement of animals coupled with manipulation of the sensory organs likely to be involved in reorienting towards the goal (e.g. Sinsch, 1987).

4.2.2 Migration by urodeles

An ability to navigate over long distances, either to a breeding site or just back to a summer home range, has been demonstrated in at least 16 North American and four European species of urodeles (Sinsch, 1991). The speed of travel varies considerably, with displaced newts *Taricha rivularis* accomplishing as much as 8 km in a year or 400 m in a single night. Migrations only occur under favourable weather conditions, which in the case of European newts *Triturus vulgaris* and *T. helveticus* means temperatures on spring nights higher than 0°C and preferably with at least some rainfall to keep humidity high (Harrison, Gittins and Slater, 1983). There is also evidence that migration routes can become traditional, in the sense that many individuals use essentially the same paths in successive years, and these corridors often correspond to attractive microhabitats such as linear depressions or streambeds when these are available (Stenhouse, 1985a; Verrell, 1987a).

The classic experiments on urodele migration were those performed many years ago on the Californian red-bellied newt *Taricha rivularis*. These newts could eventually return to their chosen breeding streams after displacements of up to 8 km, much further away than they would ever normally travel, and they primarily (but not exclusively) used their olfactory sense to achieve this end (Grant, Anderson and Twitty, 1968). Odour is also used by male red-bellied newts to home in on specific sites, notably those smelling of females, once they have entered the breeding streams (Twitty, 1955). A similar reliance on olfactory rather than visual cues has been demonstrated for plethodontid salamanders, but this is certainly not the whole of the story. Laboratory studies with the North American tiger salamander *Ambystoma tigrinum* have shown that the sun can be used for orientation, and that these animals possess an automatic time-compensator to accommodate the changing position of the sun through the day. However, it is not clear whether the sun is used directly or whether skylight polarization patterns (which in turn depend upon the position of the sun) are the critical determinants. Salamanders of this species certainly can orientate on the basis of polarization patterns, which are detected by the pineal complex situated centrally on the top of the head. There are substantial advantages for a migratory amphibian in utilizing polarized light rather than direct sunlight, because the former is accessible underwater, at twilight, and through a forest canopy provided that a patch of blue sky is visible (Adler, 1982). There is no evidence to suggest that auditory cues are used by urodeles, but there is one further sense involved in at least some species. The cave salamander *Eurycea lucifuga* and North America's eastern newt *Notophthalmus viridiscens* are disoriented by variations in the earth's magnetic field and may use magnetic information for both map and compass steps during migration (Phillips, 1986).

4.2.3 Migration by anurans

The mass movements of hundreds and sometimes thousands of frogs and toads en route to their breeding ponds in spring are surely among the most remarkable spectacles of the natural world. Walls and other barriers which appear insurmountable to human eyes are passed without obvious difficulty, but casualties are often high, whether it be from road traffic during passage across highways or from the depredations of opportunistic predators. Migration and orientation have been studied in at least 23 species of anurans, almost all of which are North American or European (Sinsch, 1991). Travelling usually, but not always, occurs after dark and in general frogs and toads move at speeds comparable with those of urodeles, covering perhaps 100–450 m per night (e.g. Van Gelder, Aarts and Staal, 1986). As with urodeles, weather conditions are important and migrations do not occur below minimum threshold temperatures (often in the range of 0–4°C) which vary somewhat between species, though in general the tolerance range of climatic variables within which movement occurs is rather broad (Sinsch, 1988).

Pioneering work on anuran migration was carried out in the 1930s at sites in England visited by common frogs *Rana temporaria* (Savage, 1961). Correlation of migratory activity with weather conditions and other factors showed that frogs tended to move upwind, or displayed positive rheotaxis when in streams or ditches, leading to the hypothesis that they were following a chemical scent to its source. Frogs of this species often select specific ponds from a range available and will return to them after physical displacement, bypassing others that appear suitable for breeding. Savage pointed out that the algal flora of ponds is often very characteristic and forms an important component of tadpole diet, and that algae produce volatile chemicals that frogs might detect and use to direct their migrations. This idea remains popular but unproven more than half a century on.

In recent years rather more work has been done on various species of bufonid toads. Animals of this genus often partake in lengthy and conspicuous migrations, thus making them particularly attractive for such studies. In North America *Bufo boreas* and *B. valliceps* have received particular attention, while in Europe the common toad *B. bufo* has stolen the limelight (e.g. Tracy and Dole, 1969; Grubb, 1973; Sinsch, 1987). *Bufo bufo* is an explosive breeder, and large numbers (hundreds or thousands) engage in mass migrations from terrestrial hibernation sites to breeding ponds during the early spring in north and central Europe. The toads travel more or less in straight lines as far as topography allows, and often through a wide variety of habitats, to reach a goal which may be up to 3 km from the start of their journey. Although the toads usually hide in vegetation during daytime, significant numbers can sometimes be found on the move in daylight. Toads will also migrate, sometimes for several consecutive years, to ponds long since destroyed (e.g. McMillan, 1963).

Investigations into the sensory basis of migratory behaviour in toads have included laboratory studies in which animals are confined and tested for orientation responses to various cues, and displacements in the field coupled with temporary deprivation of sight, hearing or olfaction, or attachment of small bar magnets. It is now clear that toads can make use of several different guidance mechanisms, which can be integrated hierarchically in a way not understood but which vary in importance depending on the ecology of the particular species. In general, olfactory and magnetic cues seem to be important to species migrating over long distances (> 1.5 km), such as *Bufo boreas* in North America and *B. bufo* in Europe; these cues can probably function as alternatives, but visual information matters much less. By contrast, toads such as *B. spinulosus* and *B. calamita*, which normally migrate only short distances, are relatively little affected if deprived of olfaction but seem highly dependent upon sight and can also use magnetic information. The visual cues are probably not connected to celestial objects but relate to fixed landmarks; in other words, these amphibians may learn the geography of their home range in a similar way to humans. Long-distance migrants such as *B. bufo* still employ vision, but primarily for short-distance piloting around obstacles. Perhaps surprisingly, acoustic cues seem to play a minor role even in species such as *B. calamita* and the chorus frog *Pseudacris triseriata*, which vocalize loudly and can often be heard over great distances. However, in these cases there are important differences between the sexes. Male *B. calamita* will pass choruses of conspecific males to return to the pond from which they were displaced, but females do respond to the vocalizations and normally move towards the nearest chorus after displacement (Sinsch, 1992a).

4.2.4 Problems about migration

Although there has been considerable progress in unravelling the mechanisms employed by amphibians to find their way (summarized in Figure 4.1), crucial questions remain unanswered. For one thing, little is known about the triggers of migration. It is common experience that in most winters there are periods, sometimes quite prolonged, in which climatic variables are compatible with migration but during which the animals do not appear. Presumably there is some type of endogenous hormone-regulated clock which generates a 'window of opportunity' for migration over a particular period each year, but this cannot be the whole of the story, because individuals of some species begin their migrations in autumn. They may hibernate en route to the pond or, for example in the case of some male (but very few female) *Rana temporaria*, may complete their migration in autumn and hibernate in the breeding ponds (e.g. Verrell and Halliday, 1985).

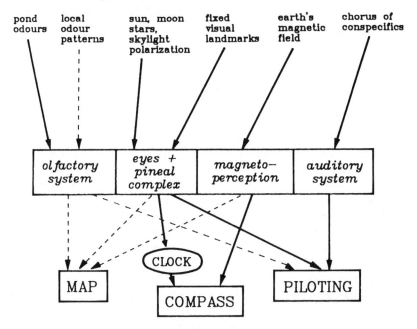

Figure 4.1 Orientation cues for amphibian migrations. (Reproduced, with permission, from Sinsch (1990a), *Ethology, Ecology and Evolution* 2, 65–79.)

It remains unclear as to how amphibians are able to orientate and return 'home' when displaced long distances into areas they are most unlikely to have experienced previously. Sinsch (1990a, 1991) questions whether we can ever be sure that the experimental subjects have not seen their release point before, but in many studies the probability of prior knowledge must have been remote. Return to a summer home-range in these circumstances is difficult to explain on the basis of olfaction. Of the known senses, use of the earth's magnetic field seems the most likely to be important. If it can be employed purely as a compass, so the animals know which way to go but not where they are (and therefore not how far they will have to travel), it may be that a 'map step' is unnecessary. The problem is understanding how the sensory system (not yet physically located in amphibians) could detect the very tiny changes in magnetic field strength, usually no more than 0.01% of the general background, that occur on the earth's surface over the distances involved. Remarkably, there is evidence that at least some urodeles can indeed respond to such tiny differences (Adler, 1982).

Finally, although olfaction has emerged as a crucial sense in the navigation of many different species of urodeles and anurans towards their breeding sites, we still know next to nothing about the chemicals involved. Deciphering this mystery poses major technical difficulties, but will be essential for a complete understanding of amphibian migration.

4.3 SEXUAL SELECTION

4.3.1 Introduction

Choosing the best possible mate is vital for the perpetuation of any individual's genes, and behaviour patterns in sexually-reproducing animals have evolved, under strong selection pressure, to achieve this end. The strategies tend to be somewhat different between the sexes. Males, producing abundant sperm at little resource cost, usually try to mate with as many females as possible; they may select the larger ones when this is a sign of fecundity, and will often attempt somehow to deprive other males of access to them. Females, producing fewer eggs at greater cost, try to select males on the basis of some estimate of their 'fitness' and may also hedge their bets by mating with several males consecutively. Amphibians, with their highly varied mating systems, provide examples of all these behaviours and their patterns of sexual selection have been very widely studied. A frequent objective in work of this kind is to try and determine whether breeding systems are dominated by male–male competition, or by female choice.

4.3.2 Sexual selection in urodeles

Plethodontid salamanders are among the most terrestrial of urodeles, and their mating success can be observed, and manipulated, in experimental terraria. There is no amplexus but courtship involves males engaging females with a characteristic 'tail straddling walk', culminating in the production of a single spermatophore for her to pick up, in a complex process which can take several hours altogether. Sexual selection in dusky salamanders (*Desmognathus* species) of eastern North America has been particularly well studied. All the evidence points to a dominance of male–male competition in the breeding systems of these animals; large males bite and chase away smaller ones, and mate with large females (which are the most fecund) when a choice is available (Houck, 1988; Verrell, 1994). Conversely, females do not compete for access to males (Verrell, 1989). In other plethodontids (genus *Plethodon*), sexual interference involving males depositing spermatophores on top of those of other males, and vigorous territorial defence of breeding areas have also been reported. Body size in males is probably the key determinant of mating success in plethodontids, but it is also known that *Desmognathus ochrophaeus* clutches sometimes have multiple paternity (e.g. Tilley and Hausman, 1976), and thus there may be some degree of female choice operating as well. Even in this situation, however, there could be sperm competition and therefore another more clandestine manifestation of male–male competition after mating.

A rather different situation pertains with the elaborate courtship displays of the aquatic-breeding newts, such as those of the European genus *Triturus*. In these urodeles there is little or no physical contact between the sexes during the breeding season but there are high levels of sexual dimorphism at this time of year, with males often bearing large dorsal crests, deeply finned tails and bright coloration – features that are absent or much reduced in females. European ponds in springtime effervesce with the activities of these newts: males actively chase females and engage them with complex, ritualistic courtship displays. The details of these routines have been thoroughly studied in several species (e.g. Halliday, 1975; Green, 1989; Sparreboom and Teunis, 1990). Males recognize conspecific females by their distinctive odour (e.g. Cogalniceanu, 1992); they initiate courtship and progress through a series of characteristic behaviours which usually include rapid tail movements (especially fanning) that convey water-borne pheromones, creeping, and spermatophore deposition on the pond floor (Figure 4.2). Females recognize conspecific males by their appearance and scent; the female role is mostly passive, though they ultimately move over the spermatophore and pick it up with their cloaca.

Sexual selection in newts varies in form between the different species. In the smooth newt *T. vulgaris*, males chase females and do not defend territories; they are sexually active mostly around dusk, and can produce up to seven spermatophores in a 24-hour period. There is no fighting but sexual interference by males, tricking other males that have initiated courtship to deposit their spermatophores and then luring the female on to one of their own, is a common tactic late in the breeding season when males greatly outnumber receptive females. Although one spermatophore contains sufficient sperms to fertilize the entire complement of a female's eggs, at least some females readily mate two or more times in close succession with either the same or different males (Verrell, 1984; Pecio, 1992). Once again, therefore, the opportunity exists for sperm competition. As with plethodontids, male smooth newts prefer to court the larger, more fecund females when given a choice (Verrell, 1986) but males do not have it all their own way. Females are looking for particular signs of fitness in their suitors; they choose those capable of putting down several spermatophores, an ability which is accurately signalled by crest size and display rate (Halliday and Verrell, 1986; Green, 1991). Thus although there is male–male competition in smooth newts, body size is not a critical factor either for sexual interference or for spermatophore production rates (Baker, 1990) and it seems likely that females are more often in a position to choose their partner than vice versa.

Other small species of *Triturus* may be similar to *vulgaris* in their mating behaviour, but the larger newts show important differences. Male *T. cristatus*, *T. marmoratus* and *T. vittatus* all establish transient territories in their breeding ponds, usually for one evening at a time, and thus maintain lek-like systems that are defended against other males by aggressive displays, chasing and even biting (Zuiderwijk and Sparreboom, 1986; Zuiderwijk, 1990). Operational sex ratios (i.e. the ratios of sexually active males to females, in contrast to the ratios of total males to females in the pond at any particular time) are highly skewed in favour of males at most sites through most of the breeding season. Males can be seen at dusk and after dark, spaced out in patches of open, shallow water around the pond margins in areas particularly suitable for courtship displays; females visit males in their territories as and when they require insemination. Among these three large species there are some differences, with *T. cristatus* males being the least aggressive and regularly engaging in sexual interference of the type described for *T. vulgaris*; *T. marmoratus* males are more aggressive, and those of *T. vittatus* highly so. Male–male competition in these species is therefore manifest in the acquisition and defence of temporary territories, but this system is very open to female choice and it has been shown that, as with *T. vulgaris*, mating success of male *T. cristatus* is correlated partly with body but mainly with crest size (Hedlund, 1990). A particular puzzle with respect to the relative importance of male–male competition and female choice in many newt species concerns female behaviour during bouts of sexual interference. Does she take account of crest height as a fitness measure and thus go with the better male, or is she usually deceived by the intruder in this situation, irrespective of his condition?

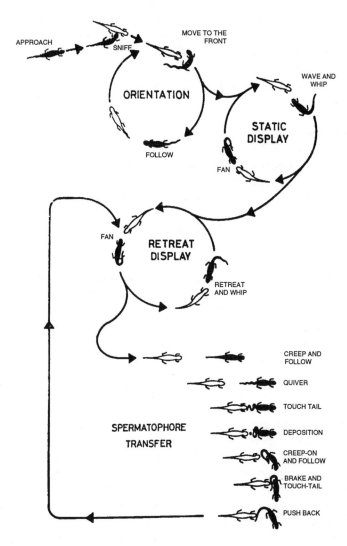

APPROACH

SNIFF

MOVE TO THE
FRONT

ORIENTATION

FOLLOW

WAVE AND
WHIP

STATIC
DISPLAY

FAN

FAN

RETREAT
DISPLAY

RETREAT
AND WHIP

CREEP AND
FOLLOW

QUIVER

TOUCH TAIL

SPERMATOPHORE
TRANSFER

DEPOSITION

CREEP-ON
AND FOLLOW

BRAKE AND
TOUCH-TAIL

PUSH BACK

Figure 4.2 Courtship routine of *Triturus vulgaris*. The male is in black. (Reproduced, with permission, from Halliday (1975), *Animal Behaviour* **23**, 291–322.)

4.3.3 Sexual selection in anurans

There are two important general differences between the mating systems of urodeles and anurans. Firstly, male frogs and toads invariably grasp females in an amplexus embrace which may last from seconds to many weeks prior to oviposition, a degree of physical contact between the sexes that is much rarer in salamanders. Secondly, male anurans usually employ vocalization during their breeding seasons whereas urodeles do not. Both of these features have consequences for the operation of sexual selection. Broadly speaking there are three fairly distinct types of breeding behaviour found among the anura, all of which are represented right across the globe in both temperate and tropical environments:

1. Explosive breeding, with all reproduction being completed in a fairly short time period (often less than two weeks). Species using this system often do so in response to environmental cues, such as a warm period in spring in temperate countries or heavy rains in the tropics.
2. Protracted breeding, with activity spread over several months.
3. Protracted breeding accompanied by male parental care, which may be thought of as a special case of (2) but which has particular implications for the operation of sexual selection.

A large number of tropical and temperate anurans operating these three patterns of reproduction have been the subjects of study by behavioural ecologists, and common themes have begun to emerge.

The common toad *Bufo bufo* is arguably one of the most successful vertebrates on earth with a distribution that includes most of Europe, central Asia and north Africa. Hundreds or thousands of adult toads descend upon their chosen breeding ponds early every spring, and engage in frantic bouts of mating and spawning that continue day and night but which are often all over within a couple of weeks. In this typically explosive breeding pattern there is furious 'scramble competition' between males for females. Males tend to arrive first, and the operational sex ratios at the ponds are usually heavily biased in their favour throughout the reproductive period, partly because males tend to mature a year younger than females (Hemelaar, 1983) and partly because females spend much less time than males at the breeding sites. Although the largest females are also the most fecund, there is little opportunity in this situation for males to choose which female they amplex and only a small proportion (perhaps 20%) of males will obtain a female at all (Davies and Halliday, 1979). The main questions are, therefore, what determines whether a male will mate successfully and do females have any say in the matter? Measurements of amplexed pairs have yielded conflicting results, with some studies finding size-assortative mating and thus a correlation between male and female size (e.g. Davies and Halliday, 1977), some finding simply that big males are the most successful (e.g. Reading and Clarke, 1983) and others finding no evidence for any kind of selection (e.g. Hoglund and Robertson, 1987). Nevertheless, fighting between males is intense (Figure 4.3)

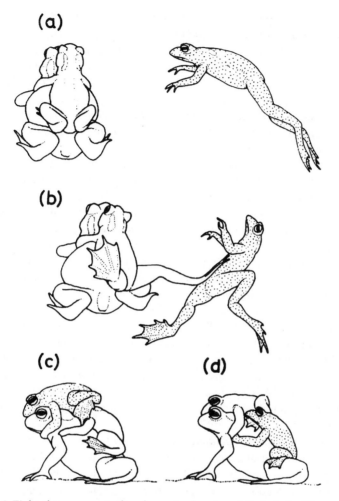

(a)

(b)

(c)   **(d)**

Figure 4.3 Fights between paired and unpaired males of *Bufo bufo*. The speckled male is the assailant. (Reproduced, with permission, from Davies and Halliday (1979), *Animal Behaviour* **27**, 1253–1267.)

and at one study site nearly 40% of the males initially paired were displaced by a competitor prior to oviposition. Male assailants could apparently judge the size of an amplexed incumbent male, and therefore whether it was worth making a serious attempt to displace him, by the pitch of his croak in response to an initial assault (Davies and Halliday, 1978). Unpaired male toads constantly search the area around the spawning site, both in the water and on land, looking either for single females or perhaps for females bearing small males that could be displaced. A high proportion (often 80–90%) of common toad females are amplexed before they reach the water's edge and substantial numbers are drowned by the combined assaults of multiple males. Although success by the largest males in this kind of conflict seems reasonable enough, it is not so easy to explain size-assortative mating since this implies that small males are more successful than large ones on small females. The advantage of this combination may be that fertilization efficiency is maximized (as indicated experimentally by Davies and Halliday, 1977), and it could arise by some form of female choice, but how it might be achieved is not clear and this remains a contentious issue.

What has become evident is that the reproductive behaviour of explosive breeding anurans is more complex, and flexible, than previously supposed and the pattern of selection may vary according to the conditions of a particular breeding season. Thus at one extreme (e.g. when an unusually prolonged winter finishes, and breeding ensues quickly without any further delay) there may simply not be enough time between the onset of breeding and spawn deposition for male–male competition to operate. In this situation pairs will be random with respect to size. However, in a more normal situation there may be some time for fighting, and big males may achieve dominance; and at the other extreme, if there is unusual protraction of the breeding season (e.g. due to interruption by a cold spell), female choice might somehow intervene and lead to size-assorted pairs by the time of oviposition. Indeed, careful scrutiny of the mating patterns of several explosive-breeder species including *B. bufo* showed that under the latter circumstances they can become remarkably similar to those of classic protracted breeders, with their relatively feeble male advertisement calls achieving unusually high prominence (Olson, Blaustein and O'Hara, 1986; Kuhn, 1993).

The natterjack toad *B. calamita* is widely but locally distributed over much of western and central Europe. Natterjacks breed in a protracted season that can extend over several months. The typical situation is one in which males, which vocalize loudly, spread themselves out around the banks of shallow ephemeral pools and call intermittently to attract females. Males defend calling territories in a lek-like breeding system in which large individuals, with the lower-pitched calls, tend to displace smaller ones; females are attracted by call intensity (and thus usually just to the nearest male) and therefore are choosing their mates after territories have been acquired by male–male competition (Arak, 1983). However, some males adopt a silent 'satellite' strategy, waiting near larger callers and trying to intercept females en route to their chosen suitor (Arak, 1988). This method often succeeds, and individual males may switch between calling and satelliting according to local circumstance. In general, therefore, there is no overwhelming advantage in large body-size for male natterjacks, but this is not always true because, as with *B. bufo*, the range of breeding behaviour is more complex than the above scenario suggests. In populations of high density and when the breeding season is relatively short, large males do dominate the amplexed pairs but not, usually, as a result of successful take-over fights (although fighting does occur). The key to success in this situation seems to be persistence; large males attend the choruses more often than small ones and obtain more females as a result (Tejedo, 1992). At another extreme, in a low density natterjack population in which the operational sex ratio often favoured females, virtually all males obtained mates irrespective of their size. Nevertheless, male strategies and success rates did vary; some called only at the start of the season, and others throughout, while some consistently called from a single site and others ('switchers') roamed between and set up calling territories in different ponds. Switchers staying active for the full season were the most successful, at no apparent cost in increased mortality between years (Denton and Beebee, 1993a).

Staying power is therefore a key to success for males of prolonged breeding season species. This is true not just for natterjacks but also for a wide range of anurans including, for example, the South American Smith frog *Hyla faber*, the males of which make nests and invite females to them (Martins, 1993); and the South African raucous toad *Bufo rangeri* (Cherry, 1993) and the South American dendrobatid *Epipedobates femoralis*, in which male success depends on the size of territory defended, which in turn relates to persistence of calling and duration of residence (Roithmuir, 1992). There is often evidence that females attempt to exert choice in these situations: in *B. rangeri* and in the African painted reed frog *Hyperolius marmoratus*, females go for males with high call repetition rates; in *Ololygon rubra*, females try to mate assortatively with males close to 80% of their own body length (e.g. Passmore, Bishop and Caithness, 1992; Bourne, 1993), though they are often thwarted by male–male competition in the form of satelliting. Other examples of female choice playing a major role in mate selection can be found in the foam-nest making tropical frog *Physalaemus pustulosus*, in which females seek the larger males with the most complex, low frequency calls (Ryan, 1980), and the American bullfrog *Rana catesbiana*, in which females again choose the biggest males, in this case because they are defending the best oviposition territories as judged by subsequent egg survival (Howard, 1978).

A third type of situation arises when males exercise parental care. This is relatively common in the tropics but unusual in temperate anurans, though the European midwife toad *Alytes obstetricans* provides one good example. In this species there is a protracted breeding season and males call intermittently from refugia on land to attract females. The eggs are wound in a string around the hind limbs of males during fertilization (which also occurs on land) and are then carried by them until hatch time when tadpoles are released into a suitable pond. Male midwife toads do not fight for females, but females compete with one another for males by displacing rivals from amplexus (Verrell and Brown, 1993). The most thorough studies of parental care have been carried out with the South American dart-poison (dendrobatid) frogs. In this complex group of rainforest inhabitants there are several species in which males defend nests until the eggs hatch, and subsequently transport the tadpoles on their backs to rearing pools. Despite this high investment, male–male competition for females is as strong as in other dendrobatids where females take on the parental care; but females also compete, not so much for access to males but to guard them after mating from access by other females and thus ensure that their own eggs receive exclusive attention (Summers, 1992).

4.3.4 An overview of sexual selection in amphibians

Perhaps unsurprisingly, mate selection by both urodeles and anurans is a complex and varied affair. Male–male competition is frequently intense and can be manifest in a number of ways. Fighting, either directly over females or in defence of a territory to be used for the attraction of females, usually produces a victor on the basis of his size. Small males may nevertheless succeed by sneaky methods such as satelliting or interference, and individuals often show behavioural flexibility between these techniques according to the circumstances of the day. In any case, size is not everything by any means; condition (having the resources and therefore the energy to keep trying for a long time) can be just as important. Although larger females are generally the most fecund, amphibian mating systems are such that there is usually little opportunity for males to choose which females they get.

Females, on the other hand, are looking for male attributes that maximize their own fitness during reproduction. In some cases such as scramble competition in explosive breeders, there is little evidence that females are able to exercise any choice at all; but in other situations they may do better, on the basis of male attributes such as overall body size, crest size in newts, and call intensity, frequency or complexity in anurans. Large male body size is probably attractive to females because it reflects juvenile growth rate and thus the speed of escape from size-dependent predation; but body size is usually only weakly correlated with age, and females will therefore not normally be in a position to judge long-term survivorship on this basis (Halliday and Verrell, 1988). Among prolonged-breeding anurans, the most persistent callers are those that have acquired the best reserves to sustain their activities: for at least some species there is now good evidence that these persistent callers lose weight, and therefore condition, at slower rates than their less successful colleagues (e.g. Cherry, 1993). Size and condition seem likely to be honest indicators of male fitness and females are therefore choosing, when they can, on the basis of 'good genes' rather than 'good taste'.

4.4 KIN RECOGNITION AND SELECTION IN TADPOLES

4.4.1 Introduction

Just as with sexual selection, kin selection is a special manifestation of the general process of natural selection. It is important because it offers an explanation of why closely-related individuals may, on occasion, behave in ways which benefit each other in an apparently altruistic manner. The underlying theory is genetical and based on the fact that, for example, siblings share 50% of their genes and so also do parents and offspring; so genes promoting individual self-sacrifice on condition that at least two siblings (or eight cousins, and so on according to the level of genetic relatedness) survive are likely to spread, and this is the process of kin selection. Anuran amphibians are in many ways ideal subjects for the study of kin selection, because their progeny (tadpoles) frequently occur in large numbers and remain for long periods in defined spaces (ponds). Most kinship studies have been carried out on anuran rather than urodele larvae because the former often aggregate into shoals and can readily be monitored both in the field and in the laboratory.

4.4.2 General aspects of tadpole behaviour

The behaviour patterns of amphibian larvae are more complex and diverse than might be anticipated on the basis of their relatively simple form. Most urodele tadpoles, and those of some anurans (such as many ranids), are rather secretive and spend much of their lives hidden in weed or silt. Other anuran larvae, particularly those distasteful to predators, often form conspicuous shoals (Figure 4.4) which, in exceptional cases, can contain hundreds of thousands of individuals (e.g. Wassersug, 1973). The reasons for shoaling may include improved thermoregulation, the stirring up of detritus particles to improve feeding efficiency, and defence from predators. Tadpole shoals are usually conspecific, though interspecific associations can occur in ways which may depend upon competitive strengths and relative palatabilities to predators (Griffths and Denton, 1992). Even when not shoaling, tadpoles move between different parts of the pond and vary in activity levels according to weather conditions and time of day (e.g. Griffiths, Getliff and Mylotte, 1988). Tadpoles also exhibit specific behavioural reactions to the proximity of predators, which are probably recognized mostly by chemical cues. According to species and circumstance this response can involve actively seeking refugia, selection of a different microhabitat or, more commonly, a general reduction in activity which, if prolonged, can stunt growth rates (e.g. Lawler, 1989; Skelly and Werner, 1990; Semlitsch and Gavasso, 1992). As metamorphosis approaches, tadpoles of some otherwise conspicuous species may become temporarily secretive, while immediately after it metamorphs may reaggregate in clumps of froglets or toadlets to conserve moisture before dispersal into the hinterland (e.g. Heinen, 1993). A few species of anurans produce tadpoles with highly specialized behaviours, such as foam-making in foam nests of the neotropical frog *Leptodactylus fuscus* (e.g. Downie, 1989).

Figure 4.4 Shoaling of *Bufo boreas* tadpoles. (Photo: A. Blaustein.)

4.4.3 Kin recognition

The first evidence that tadpoles may associate preferentially with siblings came with work by Waldman and Adler (1979) on *Bufo americanus*. Subsequent studies, mainly on North American species, have confirmed that association of anuran larvae with kin is widespread but not universal, and varies in nature even between those species which do exhibit it (reviewed by Blaustein and Waldman, 1992). Thus tadpoles of the tree frogs *Hyla regilla* and *Pseudacris crucifer*, and of two ranids (*R. pipiens* and *R. pretiosa*), showed no inclination to associate with kin in laboratory experiments (Figure 4.5). Larvae of these species tend to disperse and live singly, or occasionally form small aggregations, and there must be little opportunity for kin recognition to operate under normal circumstances in the wild. Tadpoles of the red-legged frog *Rana aurora* discriminate between kin and non-kin but are able to do so only in the early stages of larval development (usually the first 3–4 weeks of life). Little is known about the larval ecology of this species but it seems that small aggregations of young tadpoles do sometimes occur. At the other extreme, *Bufo americanus* and especially *B. boreas* larvae form large and sometimes enormous shoals which must include the progeny of many different adults. In both species there is an ability for tadpoles to recognize and associate with kin, but it is relatively weak and dictated by maternal rather than paternal factors (possibly including components of the spawn jelly). Rearing experiments showed that *B. americanus* larvae tend to associate preferentially with siblings only if reared with them early in development, and association seems to be based on avoidance of non-sibs rather than positive attraction to sibs. In natural ponds, large shoals of *B. americanus* tadpoles often split into smaller aggregations that might well represent groups of related individuals. By contrast, kin recognition in *B. boreas* tadpoles is not established early in development and is only detectable after prolonged rearing with sibs; even then, short-term exposure to non-sibs rapidly nullifies kin preference and the ability to recognize kin in this species is clearly weak.

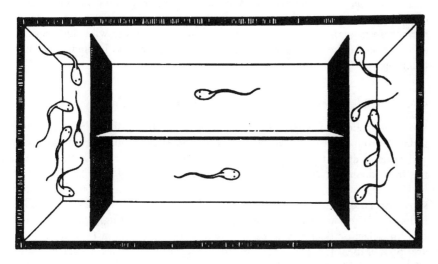

Figure 4.5 Apparatus for investigation of tadpole aggregation behaviour. (Reproduced, with permission, from Blaustein and O'Hara (1986), *Journal of Zoology (London)* **209**, 347–353.)

The strongest powers of kin recognition yet discovered are shown by larvae of wood frogs (*Rana sylvatica*) and in particular by those of Cascades frogs (*R. cascadae*). Tadpoles of the latter species recognize kin in the laboratory, using maternal and paternal factors, when reared under a variety of experimental regimes. Furthermore, sibling tadpoles marked by colour dyes associated preferentially when released into natural ponds and tadpoles taken from natural aggregations of unknown parentage reassociated together when marked and released back into ponds. *Rana cascadae* larvae frequently occur in nature in small groups of 100 tadpoles or fewer, a situation in which kin recognition might reasonably be expected. Cascades frog larvae can pick out kin groups when these are diluted 50 : 50 with non-kin, but not when the dilution is higher (25 kin : 75 non-kin), and are positively attracted towards kin rather than being repelled by non-kin. Furthermore, the preference to associate with siblings is not confined to the larval state but extends for at least 47 days after metamorphosis (Blaustein, O'Hara and Olson, 1984). Another situation in which kin recognition might be expected, and seems to happen, is between predatory and omnivorous larval morphs of the spadefoot toad *Scaphiopus bombifrons*. In this species, omnivorous larvae associate with sibs but carnivorous forms choose non-sibs; the latter explore their potential omnivorous victims with a preliminary nip, after which they release sibs but devour non-sibs, although this selection can be overruled by extreme hunger (Pfennig, Reeve and Sherman, 1993).

It seems clear that the primary mechanism of kin recognition in tadpoles is olfactory but is affected by an interplay of genetic, social and environmental factors. In general recognition has to be learnt, and in the wild this presumably happens either during embryogenesis or in the period immediately after hatching.

4.4.4 Possible benefits of kin recognition and kin selection

Why, then, should a tadpole bother to associate with its siblings? Aggregating in large groups has potential benefits for food acquisition, thermoregulation and defence against predators but many of these will accrue whether or not the individuals are closely related. In some cases it may well be that what looks like kin recognition is really just the identification of conspecifics for shoal formation (Grafen, 1990). It is also important to realize that kin recognition does not necessarily mean kin selection is operating, and conversely that kin selection does not necessarily require kin recognition. However, there are possibilities for kin selection in aggregating tadpoles (Blaustein and Waldman, 1992). These include the evolution of aposematic coloration in distasteful species, cooperative growth, direction of cannibalism towards non-kin and warning of predator attack by release of alarm substances. Kin groups of tadpoles grow faster than non-kin groups in some species such as chorus frogs *Pseudacris triseriata* (Smith, 1990) but not in others such as *Rana cascadae*, so there is no consensus on this potential benefit. However, *Rana cascadae* and *Bufo boreas* tadpoles do release substances when attacked by predators that trigger an alarm response in other conspecifics close to them (Hews and Blaustein, 1985). If kin aggregate preferentially, then they would be the main beneficiaries of such a response and kin selection could operate to favour it. Finally, if kin recognition persists long after metamorphosis, even to sexual maturity, it might provide a basis for the avoidance of inbreeding. There is some evidence, based on mitochondrial DNA analysis, that just such a mechanism may be operating in American toads *Bufo americanus* (Waldman, Rice and Honeycutt, 1992).

— 5

Population ecology

5.1 GENERAL FEATURES OF AMPHIBIAN POPULATIONS

The population dynamics of amphibians are of interest for several reasons. Firstly, there is a wide range of reproductive strategies in the group, from extreme K-selection producing very few young to extreme r-selection producing many thousands (Pianka, 1970). Another striking feature of most amphibians is their longevity compared with mammals and birds of similar sizes; many species of urodeles and anurans regularly survive into their second decade, and sometimes even longer, whereas shrews, mice and songbirds rarely reach a fifth birthday. These differences are primarily due to ectothermic rather than endothermic physiology, but also in part to the widespread use by amphibians of anti-predator toxins in defence.

The presence of amphibians, particularly at the high adult densities sometimes observed, can have a considerable impact on ecosystem function. Thus the large numbers of long-lived plethodontid salamanders that inhabit temperate forest floors in parts of North America tie up substantial quantities of resources, making them unavailable for transfer and thus retarding (and perhaps damping variations in) energy flow through the ecosystem. It has also been calculated that these salamanders consume the equivalent of more than one complete turnover of the soil fauna each year, though not all their food actually comes from this source (Hairston, 1987). Similarly, amphibians can have major effects on nutrient flows between aquatic and terrestrial habitats. At one pond in Missouri, the amount of nitrogen in tadpoles twice exceeded residual pond nitrogen during the spring months; metamorphosing *Rana pipiens*, *R. catesbiana* and *Ambystoma* species all exported 6–12 times more nitrogen from the pond than was brought in by their parents in the form of eggs, and only for one species (*Bufo americanus*) with high levels of larval mortality did nitrogen import exceed export (Seale, 1980). Amphibians may therefore reduce eutrophication, not only by net export of nitrogen but, in the case of anuran larvae, also by reducing the biomass of nitrogen-fixing blue-green algae and directly reducing primary production by feeding on all forms of algae.

5.2 METHODS EMPLOYED IN POPULATION STUDIES

Investigating amphibian populations poses a number of practical difficulties. Identification of species under study may be straightforward for adults, but is not necessarily so for eggs and larvae in areas where many amphibians breed together in mixed communities. Morphological criteria for identifying early life stages are often inadequate or even totally absent, but molecular methods are increasingly helpful in this sphere; thus, egg and embryo proteins of European newts (genus *Triturus*) can be used for identification (Veith, 1987), and spawn jelly and tadpole tail-fin proteins are diagnostic of the morphologically very similar spawn and larvae of *Bufo bufo* and *B. calamita* (Beebee, 1990).

It is also often important to mark individuals for population studies and a variety of techniques are available for this (Chapter 2). The development of skeletochronology, which by removing terminal digits permits both marking and age estimation of individual animals, has had a particular impact on studies of amphibian population dynamics (e.g. Hemelaar, 1981). It is by far the best method for estimating the age of wild amphibians (Figure 5.1), and although some uncertainties can occur due to differential reabsorption of early growth rings, it is undoubtedly more reliable than the alternatives (such as analysis of adult size distributions, or counting lobes in testes). However, growth rings are laid down most clearly in species experiencing significant growth pauses each year, usually during hibernation in temperate amphibians, and skeletochronology is therefore less satisfactory for use with tropical animals. Population sizes are best estimated by classical mark and recapture techniques, particularly the Jolly–Seber and Manly–Parr methods (e.g. Begon, 1979), though for some purposes less rigorous approaches such as counting spawn clumps or strings at breeding sites, or even counting adults directly, can prove adequate.

Finally, there is increasing interest in the genetics of wild amphibian populations and this often requires a carefully considered sampling procedure. Larvae are particularly convenient, and the taking of numbers statistically adequate for genetic studies (usually 20–40 per population) normally poses no special problems. For adults, it is better to remove small tissue samples with minimal destruction: options include toe clips again (for DNA work) or tiny quantities of blood (Wijnands, 1982). Comprehensive methodology is described in Hoelzel (1992) and by Scribner, Arntzen and Burke (1994).

5.3 POPULATION DYNAMICS

5.3.1 Long-term stability of amphibian populations

A useful prerequisite to detailed considerations of population dynamics is an assessment of how stable amphibian populations are over time. Obviously wild fluctuations in adult numbers are much more difficult to investigate than relatively stable situations, and Hairston (1987) argued that salamander populations in particular tend to be much less variable than those of comparators as diverse as lampreys, mayflies, *Paramecium*, moths, lizards, rodents and small birds. Direct counts of the terrestrial Jordan's salamander (*Plethodon jordani*) and slimy salamander (*P. glutinosus*) at two sites over 5–8 consecutive years indicated that densities usually varied by no more than two to threefold (Figure 5.2), and age structures within the populations also remained remarkably constant. Similar results were seen with the mountain salamander *Desmognathus ochrophaeus.*

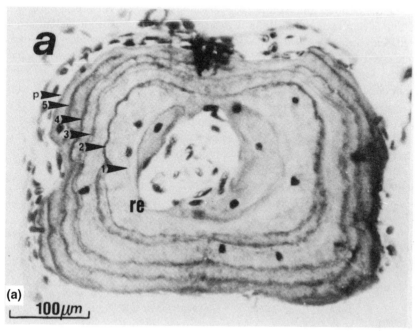

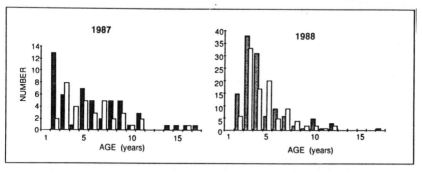

Figure 5.1 Skeletochronology of crested newts *Triturus cristatus.* (a) Stained section through phalange showing year rings; (b) age structure of newt populations in two successive years. Solid bars = males, open bars = females. (Reproduced, with permission, from Miaud, Joly and Castanet (1993), *Canadian Journal of Zoology* 71, 1874–1879.)

For pond-breeding species the data are more difficult to interpret, because they usually represent numbers coming to ponds and take no account of animals choosing to skip a breeding season. This certainly happens, and females in particular may stay away from the ponds when possible costs of reproduction (perhaps in a dry or cold spring) outweigh benefits (Bull and Shine, 1979). Pond-breeding species can nevertheless show apparent stability over substantial periods, and a large (> 8000 adults) population of common toads *Bufo bufo* at a lake in mid-Wales was stable to within 7% of its median value over four consecutive years (Gittins, 1983). However, longer studies (of which there are few) have demonstrated much bigger fluctuations. Numbers of talpid salamanders (*Ambystoma talpoideum*) at a site in South Carolina varied by about eightfold over the six years 1979–1984, but by perhaps 30-fold between 1979 and 1990 (Pechmann *et al.*, 1991). These authors also found that numbers of two other salamanders and one anuran varied by up to three orders of magnitude at the same site over this twelve year period. So although it will often be possible to study amphibian population dynamics over short periods of stability, it is important to realize that such data are likely to give an incomplete view of events unless extended over many years or even decades.

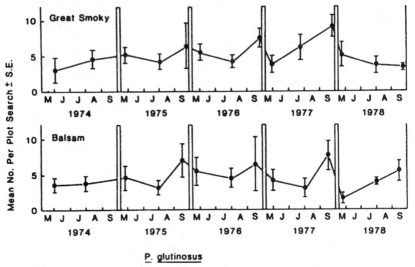

Figure 5.2 Stability of salamander *Plethodon glutinosus* populations over five years at two localities in the USA. (Reproduced with permission from Hairston (1983), *Copeia*, 1024–1035.)

5.3.2 Life-tables and mortality rates

Table 5.1 summarizes life-table information for a selection of amphibian species. The choice includes representatives of most of the major life history types, with the   significant omission of strongly *K*-selected anurans, such as dendrobatid frogs, which are abundant in the tropics but for which little information on population dynamics is available. Great care is necessary in interpreting information of this kind because there is often considerable variation within species both between different populations and in the same population over time. For some species information has been compiled from more than one study. Survivorship of adults is usually considered to be age-independent, since most individuals die before ageing effects are likely to operate, but this may not be a safe assumption in all cases. Juvenile survival rates are averages in the case of species in which the immature phase lasts longer than one year, and rates for the early year(s) are normally much lower than those of later ones. Maximum recorded ages, usually of animals held in captivity, are from Frazer (1983).

5.3.3 General aspects of amphibian population dynamics

(a) Longevity

Generation times vary widely between species, more so than indicated by the studies cited in Table 5.1. At the extremes of this variation, tropical amphibians which do not enjoy periods of dormancy may have short adult lives, as in the case of the frog *Hyla rosenbergi* which matures at one year of age and has an adult survivorship between successive years of less than 0.03 (Kluge, 1981). By contrast, species or populations living at high latitudes or altitudes, with relatively short activity periods each year, may live the longest. Examples include alpine populations of *Bufo bufo* (Hemelaar, 1986) and the Pyrenean brook salamander, *Euproctus asper*, in which many individuals survive for more than 20 years and some make their mid 30s (Montori, 1989). The age of sexual maturity varies considerably even between species living at similar latitudes and altitudes. In the neotenic *Eurycea neotenes* males mature at one year old and females at two, but this rapid growth is probably not a direct consequence of neoteny because in facultative neotenates (such as *Ambystoma tigrinum*) neotenic and normal individuals mature at the same age (one year). Conversely, terrestrial salamanders such as *Plethodon* usually mature between two and four years old. In many temperate, pond-breeding species of urodeles and anurans sexual maturity is attained after two to five years, probably depending upon (and inversely related to) population density, with male anurans often breeding for the first time an average of one year younger than females. Longevities can be compared quantitatively by calculation of mean generation times, defined as the mean age of egg-laying or giving birth. In stable populations it is calculated as $\Sigma x l_x m_x$, where l_x is the proportion of females surviving from birth to age x, and m_x is the mean number of female offspring produced per female during the time interval of which x is the midpoint. Mean generation times for a selection of North American and European urodeles ranged from 2.6 years in *Ambystoma talpoideum* to 9.8 years in *Plethodon jordani* (Hairston, 1987).

Table 5.1 Population dynamics of selected amphibians

Life history	Species	Average fecundity (eggs per clutch or year)	Hatch rate (as proportion of eggs laid)	Survival to metamorphosis (as proportion of eggs laid)	Juvenile survival per year	Survival from egg to adult	Adult survivorship between years			Maximum age recorded	
							Male	Female	Total	In wild	Anywhere (including captivity)
Neotenic	[a]Texas salamander *Eurycea neotenes*	?10–20	–	–	–	0.09	–	–	0.6	4	–
TUA, low skin toxicity	[b]Smooth newt *Triturus vulgaris*	200	0.026	0.002	0.8	0.0015	0.45	0.55	–	10	20
TUA, high skin toxicity	[c]Crested newt *Triturus cristatus*	200	–	0.025	0.22	0.005	–	–	0.49–0.78	15	25
TAA, low skin toxicity	[d]Common frog *Rana temporaria*	1330	0.96	0.05	0.15	0.0012	0.38	0.28	–	8	12
TAA, high skin toxicity	[e]Common toad *Bufo bufo*	1344	–	0.06	0.10	0.00052	0.52	0.40	–	18	36

TAA, high toxicity, strongly r-selected	Natterjack[f] *Bufo calamita*	2894	0.69	0.022						
		3813	0.84	0.003						
		4975	0.55	0.018	0.14	0.00036	0.47	0.85	–	15	15
FTU	Jordan's salamander[g] *Plethodon jordani*	10–20	–	–	–	0.119	–	0.81	25	15	–

TUA = Terrestrial urodele with aquatic larvae
TAA = Terrestrial anuran with aquatic larvae
FTU = Fully terrestrial urodele

Data sources:

[a] Bruce (1976); fecundity estimates were not made, but the related (non-neotenic) *Eurycea multiplicata* lays around 13 eggs.

[b] Bell (1977); in another study on egg mortality in the related *Triturus helveticus* the hatch rate was 0.16 (Miaud, 1993).

[c] Arntzen and Teunis (1993).

[d] Hazlewood (1969), Cooke (1975b) and Riis (1991).

[e] Gittins (1983), Gittins, Kennedy and Williams (1984) and Hemelaar (1986).

[f] Information from three populations: one in central Europe of average fecundity 3813 (Kadel, 1975), and two in Britain (Banks and Beebee, 1988; Denton and Beebee, 1993b).

[g] Hairston (1983).

(b) Terrestriality and K-selection

A general truth is that terrestrial urodeles exemplified by *Plethodon jordani* tend to have smaller clutches and lead safer, longer lives than those that are fully or partially aquatic. This may also be the case with South American anurans such as dendrobatids and Darwin's frog *Rhinoderma darwini*, in which the laying of small numbers of eggs is associated with parental care of tadpoles for various times after hatch. The young of *P. jordani* remain underground for a year after oviposition, and thus run minimal risks from predation at the life stage which is most vulnerable to it. The benefits of terrestrial life are particularly well illustrated by comparing juvenile mortality rates in various species of dusky salamanders; these decrease steadily from the fully aquatic black-bellied salamander *Desmognathus quadramaculatus*, through to the fully terrestrial pigmy salamander *D. wrighti*. Apart from minimizing the risk of death as an aquatic tadpole, terrestrial salamanders also avoid the need for potentially dangerous migrations to breeding sites. They therefore enjoy remarkable longevity, with Hairston (1987) estimating that some individuals in the populations he studied were up to 25 years old.

(c) Key factors for eggs and larvae of aquatic-breeding amphibians

Pond-breeding anurans can be more fecund, by up to an order of magnitude, than urodeles of comparable size. However, overall survival rates to metamorphosis are frequently similar between the two groups. Urodele eggs seem to be more prone to predation than those of anurans, perhaps because they are usually laid singly or in small groups, so that mortality in these early stages is relatively high. However, the active feeding behaviour of many anuran larvae makes them more vulnerable to predators than the secretive, predatory urodele tadpoles and so overall survival works out much the same. Having toxic or unpalatable tadpoles does not always improve survival rates to metamorphosis; this adaptation usually just means that different niches can be exploited (such as ponds containing predatory fish) and does not necessarily impact upon overall population dynamics. However, strongly *r*-selected species breeding in ephemeral ponds often lay more eggs and suffer a higher overall pre-metamorphic mortality in their unpredictable environments than do species breeding in permanent pools.

(d) Key factors for juveniles and adults of aquatic-breeding amphibians

Perhaps because they remain more secretive and tend to metamorphose in smaller numbers than anurans, juvenile urodeles often have substantially higher survivorships than do young frogs and toads. Once maturity is reached, survivorship is generally good for most pond-breeding amphibians though not normally as high as in fully terrestrial species. In some species survivorship is similar for both sexes, while in others either males or females regularly do better. Male anurans of protracted-breeding species may have substantially lower survival rates than females because they remain for much longer periods at the breeding ponds in situations of relatively high risk. By contrast, females of explosive-breeding anurans often fare worse than males because scramble competition can drown them. Comparable arguments can be made for newts: male *Triturus vulgaris* are conspicuous and very active in the breeding ponds, continuously darting around in their search for females, and may suffer extra predation as a result. By contrast, territorial male *T. cristatus* are much less conspicuous and not obviously at greater risk than females.

Toxic secretions may be associated with greater longevity in adult amphibians. Thus crested newts in the wild can live up to 50% longer than smooth newts, and common toads may live 20–30% longer on average than common frogs in the same habitats. The data do not suggest that *Bufo calamita* lives significantly longer than *B. bufo*, and the former species must rely mostly on its greater fecundity to cope with its less predictable breeding habitat. However, the European yellow-bellied toad *Bombina variegata* frequently breeds in puddles and other transient water bodies but females produce only a few hundred eggs each season. In this species the highly toxic skin secretions of the adult toads probably do make a critical difference; many live more than 10 years and some more than 20, so populations can persist even when breeding is only occasionally successful (Plytycz and Bigaj, 1993).

5.4 DETERMINANTS OF POPULATION DYNAMICS

5.4.1 Eggs and egg-laying

The positioning and timing of oviposition as well as the quantity and the quality of the zygotes are of critical importance to survival. Most amphibians demonstrate species-specific preferences with respect to oviposition sites; common frogs (*Rana temporaria*), for example, select shallow water exposed to sunlight, whereas common toads (*Bufo bufo*) prefer deeper water with aquatic vegetation to wrap their spawn strings around (Cooke, 1975a). The frog strategy maximizes development rate at the cool time of year when this species breeds, whereas that of the toad presumably reduces predation and desiccation risks. Amphibians also make decisions about oviposition timing: common frogs sometimes wait around in ponds for weeks after the inward migration before the onset of spawning (e.g. Savage, 1961). This may be adaptive in some way, or might just be due to a requirement for an 'accumulated temperature effect' to ensure full maturation of eggs and sperms (Beattie, 1985). Ova of the smooth newt *Triturus vulgaris* that are laid near the midpoint of the breeding season survive better than early or late ones and this timing is probably adaptive (Bell and Lawton, 1975). Anurans with protracted breeding that use temporary ponds, such as the natterjack *Bufo calamita*, can recruit primary oocytes at any time through the activity season and females within a population often show substantial asynchrony in ovarian development (Silverin and Andren, 1992). Many such species can also hedge their bets by laying multiple egg masses at different times, many weeks apart, depending on the availability of breeding sites. In general it is probably advantageous to breed as early as possible: early cohorts of the Pine Barrens tree frog *Hyla andersoni* grew better and metamorphosed at larger sizes than later ones, although early tadpoles were more susceptible to competition from other species (Morin, Lawler and Johnson, 1990).

The sizes as well as the numbers of eggs laid often increase as a function of female body size in amphibians, but the relationships can be complex and variable. Both parameters increase with female size and age in *Bufo calamita*, generating an exponential increase in reproductive investment (Banks and Beebee, 1986), but in *Rana temporaria* egg size is negatively correlated with fecundity after adjusting for body size effects (Cummins, 1986a). Large eggs can be adaptive because tadpoles hatching from them grow faster, and metamorphose earlier, than those from small eggs; they may be especially advantageous where the time available for development is short, such as at high altitudes or in temporary ponds. Slower growth, however, often produces larger individuals at metamorphosis. An alternative strategy in unpredictable environments, therefore, is to lay clumps with varying sizes of eggs and thus hope to maximize reproductive fitness whatever the circumstances of the season (Crump, 1981). Clearly there must be complex trade-offs between fecundity and egg size which probably vary between populations according to predominant local conditions.

Initial viability of eggs is usually high and the same is true of fertilization efficiency, though examples of substantial failure in both are known, and in the crested newt *Triturus cristatus* and its close relatives the genetic load due to the chromosome 1 heterozygosity requirement causes an early (pre-hatch) loss of 50% of all progeny. Newt ova, often laid singly and wrapped in the leaves of pondweeds, are particularly vulnerable to predators. Caddis fly (Trichoptera) larvae can be major egg consumers (Bell and Lawton, 1975) and Miaud (1993) noted severe attacks by dytiscid water beetles and adult newts; protection of eggs in predator-proof cages increased survivorship fivefold, from 0.16 to 0.79. Oophagy seems to be common in urodeles, and may often be the single major cause of egg mortality. The large clumps and long strings of spawn characteristic of many anurans are relatively resilient to predation, but water birds such as moorhens *Gallinula chloropus*, flatworms (Turbellaria) and newts sometimes cause substantial mortality of eggs of palatable species like *Rana temporaria*; even human predation is occasionally significant for this frog, with up to 20% of one population's output being taken by people at a site in central England (Cooke, 1985). The smaller egg clumps of neotropical hylid frogs are also vulnerable to numerous predators, including snakes. Unpalatable eggs, such as those produced by *Bufo* species, are left alone by most animals but there are exceptions: *B. calamita* spawn, for example, is occasionally devoured by conspecific tadpoles and also by those of other sympatric anurans such as the spadefoot toad *Pelobates cultripes* and the parsley frog *Pelodytes punctatus* (Tejedo, 1991). Oophagy is well developed in the larvae of some tropical anurans, particularly those living in bromeliad pools, where they may constitute the major cause of egg mortality in these highly specialized environments. Desiccation (caused by rapidly lowering water levels), unusually low temperatures, excess ultraviolet light and acidic water flushes all cause deaths of amphibian eggs and, when not immediately lethal, these agents may also predispose ova to infection by fungi such as *Saprolegnia*. Mycelia of this pathogen readily spread to overwhelm otherwise healthy embryos and can then destroy entire spawn clumps or strings. Fungi have been implicated in the deaths of tropical as well as temperate anuran ova (Villa, 1979).

5.4.2 Larvae

Urodele tadpoles are carnivorous throughout life, devouring a wide range of small invertebrates (e.g. Kuzmin, 1991) and, when large, other amphibian tadpoles including conspecifics. Those of some species show a distinct ontogenetic shift from passive hunting to more active foraging, and in the tiger salamander *Ambystoma tigrinum* specialized cannibalistic morphs can arise under conditions of high larval population density. Urodele larvae are, for the most part, solitary animals that rely on seclusion in dense aquatic vegetation to protect them from predators but are nevertheless devoured by a wide range of aquatic invertebrates including dytiscid water beetles, backswimmer bugs (e.g. *Notonecta* species) and odonate larvae (e.g. McCormick and Polis, 1982), as well as vertebrates such as fish, water birds, adult amphibians and amphibious reptiles. Even the tadpoles of newts such as *Triturus cristatus*, which have powerful skin toxins as adults, are highly vulnerable to fish predation.

Anuran larvae, by contrast, are usually omnivorous and obtain most of their food by filter-feeding in the water column or by grazing on epiphyton or detritus (e.g. Harrison, 1987; Diaz-Paniagua, 1989). Large amounts of time are spent actively feeding and apparently exposed to predators (which are generally similar to those of urodele larvae), but the trade-off is that rapid growth minimizes cumulative predation risks, which decline as a function of tadpole size (e.g. Cronin and Travis, 1986). Unpalatability of larvae is better-developed in some anurans than it is in urodeles, and *Bufo* tadpoles (for example) are usually, but not always, left alone by vertebrate predators (e.g. Peterson and Blaustein, 1991). Invertebrate predators, which have piercing and sucking mouthparts, are rarely deterred by skin chemicals (e.g. Kruse, 1983) and toad larvae often suffer high mortality from this source. Anurans, too, have their specialist cannibals: tadpoles of the plains spadefoot toad (*Scaphiopus bombifrons*) suffer major mortality from premature pond desiccation, but the production of a rapidly-growing, cannibalistic larval morph permits successful metamorphosis of the few at the direct expense of the rest.

Premature pond desiccation is a further and often major cause of mortality for the larvae of many species of amphibians. Wilbur and Collins (1973) suggested that development time reflects an evolutionary trade-off between desiccation risk and the survival advantages of large body-size, attained after longer growth, at metamorphosis. In this model, metamorphosis required the attainment of a minimum body-size (the 'metamorphosis threshold'), whereas in an alternative proposed by Smith-Gill and Berven (1979) the stage of development was considered more important than size in a differentiation-based model of the timing of metamorphosis. Subsequent studies have largely favoured the former but are often not completely consistent with either scheme (e.g. Alford and Harris, 1988). Werner (1986) further refined models of metamorphosis to take account of important interspecific differences, explaining for example the smaller size of *Bufo* metamorphs, compared with those of *Rana* and *Hyla* species of similar adult sizes, on the basis that terrestrial life was relatively safer for bufonids because of their skin toxins and they might therefore maximize fitness by leaving the aquatic habitat earlier than the others. Changes of growth rate (dependent on food supply) at any time during development might affect both age and size at metamorphosis, but investigations of *Pseudacris crucifer* supported a 'fixed rate' model in which only early growth rates, prior to a specific ontogenetic stage, can exert such effects; after that stage, the time to metamorphosis is fixed irrespective of food supply (Hensley, 1993). In tree frogs *Hyla gratiosa* and *H. cinerea*, too, food levels only affected the timing of metamorphosis when varied during early growth stages, though at later times they had a substantial effect on the size of the metamorphs (Leips and Travis, 1994). Nevertheless in at least some cases tadpoles can respond adaptively to resource variation, and in temporary ponds, where resources of both food and water diminish with time, tadpoles of Couch's spadefoot toad (*Scaphiopus couchii*) are able to metamorphose earlier than they otherwise would (Newman, 1989).

At the high densities sometimes attained by amphibian tadpoles, both intra- and interspecific competition can have substantial effects on growth rates and thus, indirectly, on survival to metamorphosis. *Ambystoma* salamander larvae grow slower at high densities and may also experience reduced survival (e.g. Hairston, 1987; Buskirk and Smith, 1991). The same is true for almost all anuran species that have been tested, and size at metamorphosis is also usually inversely related to tadpole density (e.g. Travis, 1984). Marbled salamander *Ambystoma opacum* tadpoles reared at low densities metamorphosed with higher lipid reserves, showed nearly fourfold better survival to sexual maturity, matured a year earlier and were more fecund than those reared at high densities, all of which were in the ranges found in natural ponds (Scott, 1994). Similarly, *Bufo bufo* tadpoles reared at low density metamorphosed nearly 50% larger than those at high densities, and were still 18% heavier (and thus probably fitter) at the onset of their first hibernation (Goater, 1994).

Finally, there is undoubtedly also a genetic aspect to larval development; variability of life-history traits is at least partly heritable, for example with respect to paedomorphosis in *Ambystoma talpoideum*, between montane and lowland populations of wood frogs *Rana sylvatica*, and between families of the tree frog *Hyla crucifer* (Semlitsch and Wilbur, 1989; Berven, 1987; Travis, Emerson and Blouin, 1987).

5.4.3 Juveniles

The time between metamorphosis and sexual maturity remains the least documented part of amphibian population dynamics. Obviously it is a period of rapid growth, but in many species much of juvenile life seems to be spent in seclusion and attempts at study are often confounded by the fundamental problem of finding the animals. Large numbers of newly metamorphosed anurans are sometimes killed by stochastic factors, such as desiccation around the pond banks in hot weather, and by a wide range of vertebrate and invertebrate predators. What happens between this time and the reappearance of sexually mature individuals at the pond one or more years later remains poorly understood, but it seems likely that heavy selection operates during juvenile life. In *Bufo boreas*, genotype frequencies change as the toads grow, indicating that survival is not genetically random (Samollow, 1980). *Bufo bufo* metamorphs from tadpoles reared at low densities grew and survived much better than those grown as larvae at high densities, indicating that adverse conditions experienced by larvae may not be recoverable in later life (Goater, 1994). Large body size and rapid growth rates seem likely to be important determinants of fitness, minimizing risks from predation and accidental deaths due to environmental stochasticity. Parasitism can also be significant; juvenile common toads *B. bufo*, for example, are susceptible to infection by lung-dwelling nematodes which can be lethal (Goater and Ward, 1992).

5.4.4 Adult life

Adult amphibians consume a wide range of (mainly) invertebrate prey, and diets have been examined in detail for many species. Almost anything in the appropriate size range, and capable of being caught, will usually be eaten; distasteful species, such as the larvae of some lepidopterans, are normally left alone but in general the diet reflects whatever is available in the habitat, although some species do specialize on particular groups such as ants.

Life for an adult amphibian is relatively safe, with survivorship between years commonly exceeding 0.5. In the case of pond-breeding species, much of the mortality that does occur happens en route to, or during residence at, the breeding site; breeding behaviour often further exhausts reserves that were in any case depleted during hibernation, and also renders the animals particularly vulnerable to predators. Outside the breeding season, hunting on land is usually carried out after dark, within a small home range, and often on only a small fraction of the nights in any particular year. The opportunities for predators of amphibians are therefore limited, and frequently made more so by an ability for rapid escape (leaping), cryptic coloration or toxic skin secretions. Other amphibians remain in or near ponds and streams during their adult lives, and often use them as refugia by diving in or seeking cover in the bottom sediments whenever threats arise. Nevertheless, most adult mortality of amphibians is probably caused by predation and the range of animals that will take them when they can is a large one. The great majority are vertebrates, including many species of bird, especially waders, and corvids which are efficient at disembowelling amphibians with toxic defences and leaving the skin uneaten; mammals of all sizes, from lions down to shrews (e.g. Brodie and Formanowicz, 1981); many reptiles, but especially snakes; predatory fish such as pike (*Esox* species); and, of course, other amphibians. Frog-eating tropical bats have received particular attention because they can select males of prey species on the basis of their advertisement call characters (Tuttle and Ryan, 1981). Occasionally even invertebrates take their toll, and on Puerto Rica giant crab spiders (*Olios* species) are important predators of all but the largest leptodactylid frogs *Eleutherodactylus coqui* (Formanowicz *et al.*, 1981) while as many as 8% of adult common toads (*Bufo bufo*) in some populations fall victim to flesh-eating larvae of the fly *Lucilia bufonivora* (Strijbosch, 1980b). The European medicinal leech, long thought of as primarily a mammalian parasite, can also survive mainly on frogs and newts (Wilkin and Scofield, 1990).

Infection by pathogens and parasites may also be important contributors to amphibian mortality statistics, but although many instances of infected lesions and parasite burdens have been reported, little quantitative work has been done in the field. However, the parasitic polystomatid helminth *Pseudodiplorchis americanus* can have serious consequences for its anuran host in the deserts of Arizona. Larvae of this parasite infect adult spadefoot toads *Scaphiopus couchii* during the very brief period each year when they enter the water to breed; the larvae migrate to the bladder and mature there during the rest of the year while awaiting the next desert rains to trigger another toad breeding season (Tinsley and Jackson, 1988). Prevalence of infection is regularly between 50 and 100%, and the burden for individual toads is often high enough to cause severe depletion of reserves during hibernation; it is very likely that some toads perish for this reason (Tocque, 1993).

Of course not all amphibian mortality is due to predation or infection; environmental fluctuations can also have an impact. Heat, extreme cold, drought and flooding have all been documented as killing substantial numbers of amphibians from time to time. Smith (1964) recounted nearly 50% of smooth newts *Triturus vulgaris* in one population dying during a particularly cold winter because their shelter was not sufficiently frost-proof, and frogs can suffer mass mortalities from anoxia when hibernating underwater beneath ice for unusually long periods. Whether such mortality factors are age-dependent remains unclear, and though immature animals are often thought to be most at risk, Smith noted that juvenile smooth newts seemed to survive the extreme cold better than adults did. Most of the time the refugia chosen or made by amphibians are very effective at protecting them from the vagaries of climate; burrowing, in particular, reduces water loss in arid habitats and allows animals to adjust their position during hibernation in response to temperature fluctuations above ground (Hoffman and Katz, 1989; Van Gelder *et al.*, 1986b).

Features relevant to adult survivorship are naturally selected on the basis of optimizing fitness, with a trade-off between resource allocation into somatic growth or into germ cells. Since fecundity in females and mating success in males, as well as relative security from predation in both sexes, are often related to body size it usually pays an amphibian to be large; growth is normally indeterminate and asymptotic, and the key to success is probably to achieve as big a body as possible before sexual maturity; after that point, somatic growth slows down dramatically and individuals that are small at that stage do not catch up with their larger siblings. Animals choosing not to breed in a particular year, however, can grow faster than individuals of the same size class that do reproduce (e.g. Ryser, 1989). In low density anuran populations there can be simple positive correlations between fecundity, body size and age but when the numbers of adults become high these relationships may break down (Figure 5.3); individuals then vary much more in their growth rates, presumably because of competition, and some old individuals remain small and of low fecundity. Newts, too, can attain population densities high enough to affect the condition of individuals adversely as shown by translocation experiments with *Notophthalmus viridiscens* (Gill, 1979). In extreme cases of very high population density, prey may even become scarce enough to cause substantial mortality of adult amphibians by starvation (Tyler, 1976).

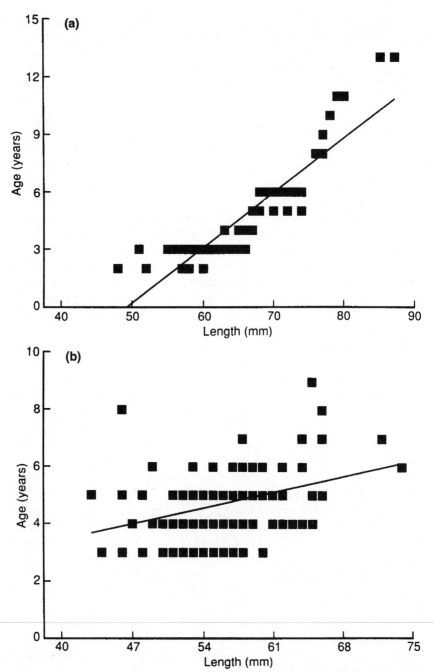

Figure 5.3 Age–size relationships of *Bufo calamita*: (a) site with low population density; (b) site with high population density. (Reproduced, with permission, from Denton and Beebee (1993b), *Journal of Zoology (London)* **229**, 105–119.)

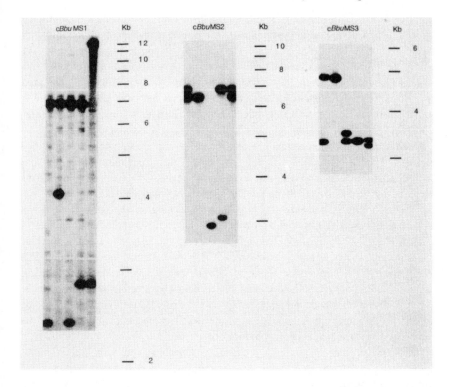

Figure 5.4 Single-locus DNA fingerprints of 5 *Bufo bufo* individuals with three separate probes. (Reproduced, with permission, from Scribner, Arntzen and Burke (1994), *Molecular Biology and Evolution* **11**, 737–748, published by the University of Chicago Press.)

5.5 POPULATION GENETICS

In addition to the multiple extrinsic factors that affect the dynamics of amphibian populations, the intrinsic effects of genetic composition must also be considered. Genetic studies provide information on levels of variation within and between populations, and make possible the assessment of gene flow (i.e. whether populations are continuous or partially separated by ecological barriers) and risks of inbreeding depression. Molecular methods, especially allozyme analysis but increasingly also DNA-based techniques (Figure 5.4), allow the critical genetic characteristics of populations to be quantified and compared. These are P, the proportion of loci that are polymorphic, and H, the proportion of heterozygous individuals. A polymorphic locus is one in which the most abundant allele occurs at less than 99% frequency; H = $\Sigma(N_h/N)/n$, where N_h = the number of heterozygous individuals at a particular locus, N = total number of individuals in the sample, and n = total number of loci studied, including monomorphic ones. With this information it is possible to test whether populations are in Hardy–Weinberg equilibrium (i.e. whether genome frequencies correspond to those predicted by theory for a panmictic group) or whether the 'population' deviates significantly from this ideal, as it will for example if the population is really a set of partially isolated subpopulations. Comparisons between populations can be made using various statistical procedures, including calculation of coefficients of gene differentiation (GST), fixation indices (F statistics) and genetic distances (Nei, 1987).

There are sound theoretical reasons for believing that in general large populations are able to maintain higher genetic variation than small ones, and that both the process of becoming small and staying small for several generations can lead by random drift to the loss of alleles by chance rather than by selection. The long-term viability of a population will be greater if genetic diversity is high, because the chances of genotypes being present that can withstand environmental change or fluctuation are correspondingly elevated. By way of example, a study of song sparrows in Canada showed that inbred animals following a natural population bottleneck were heavily selected against during the subsequent recovery (Keller *et al.*, 1994). Since mutation rates are usually very low in comparison with the speed at which populations change size, once diversity has been lost it takes a long time (in the absence of immigration) to recover even if the population subsequently expands again.

Allozyme studies of common frogs *Rana temporaria* at a site in Germany showed that populations were effectively separated from one another and had become genetically impoverished as a result of isolation by motorways and railways (Reh and Seitz, 1990). More investigations of this kind would be very valuable, but others have proved interesting for a different reason. An underlying assumption of the genetic methods is that the allelic variation observed is selectively neutral, but there is evidence that this may not always be true and that genetic analyses might also give clues to the molecular basis of selective processes. Thus in the North American newt *Notophthalmus viridiscens*, variations in four out of five loci correlated with environmental features' and in the green toad *Bufo viridis*, two particular alleles shifted in frequency in concert with a rainfall gradient across the area in which the populations lived. This subject is comprehensively reviewed by Nevo and Beiles (1991) in an analysis of genetic studies on 123 urodele and 66 anuran species from which it is apparent that genetic diversity (mainly H) is significantly related to life style and habitat type. Amphibians in general have high levels of H (averaging 0.073) compared with other vertebrates, and this ranges from its highest values in tropical, terrestrial or arboreal habitats down to the lowest in temperate, aquatic or subterranean ones. Variability is therefore greatest in the most heterogeneous and/or unstable habitats, and minimal in homogeneous, predictable ones. Average P and H values for a range of amphibian families are summarized in Table 5.2.

Table 5.2 Genetic diversity of amphibians

FAMILY	Number of species	Average *P*	Average *H*
Plethodontidae	102	0.248	0.077
Salamandridae	13	0.244	0.058
Proteidae	3	0.060	0.017
Ambystomidae	4	0.025	0.009
Cryptobranchidae	1	0.015	0.006
Discoglossidae	3	0.464	0.140
Rhacophoridae	5	0.464	0.123
Sooglossidae	3	0.322	0.106
Bufonidae	9	0.366	0.105
Hylidae	15	0.261	0.051
Ranidae	22	0.233	0.075
Ascaphidae	1	0.166	0.048
Pelobatidae	7	0.299	0.044

Source: Nevo and Beiles (1991).

5.6 METAPOPULATIONS AND THE COLONIZATION OF NEW HABITAT PATCHES

5.6.1 Metapopulations

Populations are not always discrete but can be variable in both space and time. Metapopulation theory was developed to accommodate situations in which patches of habitat are sometimes occupied and sometimes not, depending upon both demographic and environmental stochasticity. In such situations it is essential that animals can move freely between habitat patches, so that declines or even temporary local extinctions can be reversed by occasional immigration. Amphibians, particularly pond-breeding species, may behave as metapopulations and several studies have been carried out to examine this idea. In a series of ponds used by red spotted newts *Notophthalmus viridiscens*, population turnover was fast because of a short adult life-expectancy; many local populations were not self-sustaining, but were boosted each year by newly mature adult colonizers originating in ponds where recruitment was high. Site fidelity of adults was high irrespective of whether the pond they had chosen to breed in was a successful site for reproduction, but all ponds were occupied every year because of the constant high output of young newts from the successful sites (Gill, 1978). By contrast, crested newts *Triturus cristatus* using a series of adjacent ponds in France were much longer lived as adults, and did not show strong breeding site fidelity; ponds were occasionally abandoned by adults moving to different, nearby pools (Miaud, Joly and Castanet, 1993). In this situation not all ponds attracted young breeders, and emptiness of a habitat patch apparently reflected choices by living animals rather than deaths of previous inhabitants.

Another interesting situation is that of the pool frog *Rana lessonae* near the northerly edge of its range in Sweden. These anurans occupy a series of ponds, separated by various distances, and have suffered numerous local extinctions in recent decades. Sjögren (1991) showed that the probability of extinction was increased either by obvious factors (destruction of ponds) or just by distance from the nearest extant site. Where the frogs died out it was not because of inbreeding depression in small populations, but because stochastically-acting factors (mostly fish predation coupled with low reproductive success rates) could reduce the population to zero unless there was occasional immigration from a nearby site which, by chance, had enjoyed a recent successful breeding year. Low genetic variability of the isolated Swedish pool frogs was apparently not of immediate significance, since fertilization rate and viability of spawn was high. Connectivity of habitat patches is obviously essential for such metapopulations to persist, and the value of suitable habitat (especially woodland) between ponds as well as the maintenance of ponds for long time periods was demonstrated in a study of amphibian species diversity in the Netherlands (Laan and Verboom, 1990). Older ponds sustained the greatest number of species, and colonization of new ponds was strongly correlated with the local abundance of the species involved.

A most unusual kind of metapopulation structure was found in four neighbouring breeding sites of the natterjack toad *Bufo calamita* in Germany. Adults, particularly males, showed strong breeding site fidelity and a spatial metapopulation structure evidently existed; three out of the four breeding sites had very low reproductive success, and were maintained by juvenile output from the fourth. However, there was also evidence of temporally separated metapopulations; three distinct waves of breeding animals used the ponds at different times of year, and were more distinct from each other genetically than were the four spatially separated groups. The latest-breeding of the three temporally separated groups enjoyed little reproductive success and was presumably sustained by recruitment from the earlier ones (Sinsch, 1992b).

5.6.2 Colonization of new ponds by amphibians

Because many adult amphibians remain faithful throughout their lives to the breeding site they choose when first sexually mature, the general belief is that colonization of new ponds is carried out primarily by juveniles. However, this fidelity may sometimes be overstated. As shown by Miaud, Joly and Castanet

(1993), crested newts *Triturus cristatus* can switch their breeding site allegiance when mature. Arntzen and Teunis (1993) considered that rapid colo-

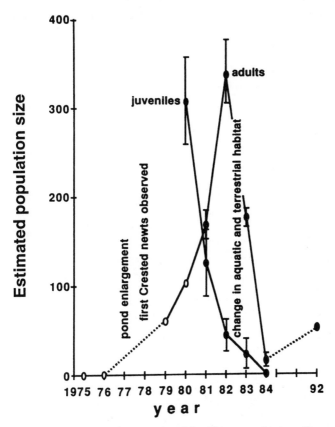

Figure 5.5 Colonization of a new pond by *Triturus cristatus*. (Reproduced, with permission, from Arntzen and Teunis (1993), *Herpetological Journal* 3, 99–110.)

nization of a new pond in northern France by this species (Figure 5.5) was carried out at least partly by old adults.

Nevertheless, juveniles are probably the main colonizers in most situations. Thus the natterjack toad *Bufo calamita* is well known as a pioneer species that rapidly establishes itself in newly created ponds during road construction or other building works, and very large numbers of juveniles are produced during this pioneering phase (Boomsma and Arntzen, 1985). A balance, not well understood, must exist between a tendency for juveniles to spread into new areas and an inclination to return when mature to the natal pond. In one particular study, 81% of male common toads *Bufo bufo* breeding for the first time, having been marked as toadlets, turned up at their natal pond despite the availability of other ponds, used by other common toads, in the same general vicinity (Reading, Loman and Madsen, 1991). In this instance, distances of more than 300 m between ponds markedly separated populations, and distances of more than 800 m were enough to ensure complete genetic isolation.

— 6 —

Community ecology

6.1 THE IMPORTANCE OF COMMUNITY STUDIES

Populations exist in a matrix of other species struggling to survive in the same habitat patch. Unravelling this complex web of interactions is the business of community ecology. **Communities** are usually defined as assemblages of organisms without taxonomic restriction, whereas **guilds** are groups of species, usually taxonomically related, which are thought to have similar roles in the community and thus exploit the same resources in the same way. Both concepts are important in community ecology, and a substantial body of ecological theory has developed around them (e.g. Hairston, 1987). Broadly speaking, communities are thought to be structured by four major forces: competition, predation, specialization and environmental stochasticity. Competition can be further subdivided into direct, exploitative competition for food, space or some other resource, and interference competition in which individuals indirectly affect the abilities of others to obtain a critical resource (by, for example, aggression).

The relative importances of these forces in nature remain the subject of contentious debate, and in many situations some or all of them are likely to interact; it is well known, for example, that the superimposition of predation often markedly affects the outcome of competition between sympatric species. A further complication when contemplating work with pond or stream-breeding amphibians is that they are components of two distinctly separate communities at different times in their life cycles. Many studies have concentrated on guilds of related species, partly because this makes an inherently complex field easier to cope with, but also because it is within guilds that competition is most likely to occur. However, the more difficult problem of studying interactions over a range of trophic levels is receiving increasing experimental attention as indeed it must if ecosystems are ever to be properly understood.

6.2 METHODS IN COMMUNITY ECOLOGY

Studies on community ecology tend to be of two types: observational ones, in which detailed measurements lead to inferences about how communities function, and experimental work in which hypotheses can be tested directly. Observational studies often focus on niche breadth and overlap, and seek to determine the extent of these quantitatively for niche dimensions such as food and space. There are several methods for quantifying niche parameters that have proved useful in amphibian studies (e.g. indices B, C, B' and L; Levins, 1968; Schoener, 1970; Hurlbert, 1978). Experimental studies include manipulating the densities of one or more community components within enclosures and monitoring effects on the other species present (e.g. Hairston, 1987), and the use of replicated artificial ponds or laboratory tanks in which community composition can, at least to some extent, be controlled by the experimenter (e.g. Rowe and Dunson, 1994).   Although the experimental approach is in principle more powerful than the observational one, the high complexity of biological communities limits its applicability more than in perhaps any other area of science. Small differences in the starting conditions of experimental ponds, for example, can lead quite quickly (within the time course of an experiment) to substantial divergence between what started out as identical replicas. Despite this problem, some major advances in amphibian ecology over the past 20 years have stemmed from the successful application of just such experimental approaches.

6.3 OBSERVATIONAL STUDIES OF COMMUNITY ECOLOGY

6.3.1 Communities of adult amphibians

It is in the tropical rainforests that amphibians, particularly anurans, attain their greatest diversity and therefore it is in these habitats that amphibian communities are most complex. Major studies have been carried out at Santa Celia in Ecuador, where 81 species occur, and in the Amazonian rainforests of Peru and Brazil, with 53 and 37 species respectively (reviewed in Duellman and Trueb, 1986). In the primary forest at Santa Celia the most striking niche partitioning was of vertical space, though the terrestrial communities also partitioned horizontally between swamp and leaf-litter, and temporally between nocturnal and diurnal activity. Bush and tree-dwelling species, at the higher levels of stratification, were entirely nocturnal. Thirty-nine species were ground-living (including seven categorized as diurnal, 13 as nocturnal and nine as both); 46 were bush-dwellers, and 24 lived in trees more than 1.5 m above ground. There was little partitioning of macrohabitat defined as primary or secondary forest and forest clearings; 58 of the 69 species inhabiting primary forest were also among the 68 species using secondary forest, while 18 primary forest and 25 secondary forest species were among the 26 species seen in clearings. Other niche dimensions were also partitioned, though less dramatically than vertical space; thus food differed between species on the basis of both type and size. Seventeen species specialized on ants, one on termites and two on other frogs, while the remaining members of the community were sit-and-wait predators taking a wide variety of prey items. Sit-and-wait predators were usually cryptically coloured, while the major alternative strategy of active foraging was often associated with toxic secretions and aposematic coloration. There were also many different reproductive strategies, involving a range of distinctive oviposition sites. While breeding, tropical anurans also partition the acoustic environment and vocalization is of primary importance in species recognition.

Other studies have concentrated on particular parts of the tropical forest ecosystem. Anurans living in leaf-litter partitioned along moisture gradients (probably as a result of specialization) where the environment was heterogeneous in Panama but no such segregation was seen in the more homogeneous conditions of Peru and Gabon. One of the species least resistant to desiccation, *Dendrobates auratus*, lived in damp ravines in Panama and was absent from the drier ridges, but foraged in dry areas on Taboga island where anuran competitors were absent, suggesting that in this case at least there was more to niche separation than specialization. Dietary overlaps between species were often high but diminished when food was less abundant, again suggesting that competition between some species was operating at least some of the time (Toft, 1980). In a South Indian rainforest with 52 species of amphibian and reptile, distinctive guilds of terrestrial species were identified. There were no general relationships between niche breadth and general habitat type (e.g. terrestrial or arboreal) but altitudinal partitioning, corresponding to changes in forest types, was very marked in this fauna (Inger *et al.*, 1987).

There have also been numerous observational studies of niche partitioning in temperate adult amphibian communities, of which the European newts (genus *Triturus*) constitute a good example. There is widespread sympatry between several of these newts, especially *T. vulgaris*, *T. helveticus*, *T. alpestris* and *T. cristatus*, and it is not uncommon to find ponds in which three or even all four species breed together in springtime (e.g. Glandt, 1980). *Triturus vulgaris* and *T. helveticus* are very similar in morphology and behaviour, and demonstrate high niche overlaps in all dimensions studied, including seasonal use of the breeding site, microhabitat selection within the pond and food intake (Griffiths, 1986, 1987). The prospects for competition between these species would seem to be high but may not often be realized if population sizes are limited by other factors such as predation. By contrast, *T. cristatus* is much larger than both the former species but frequently occurs together with *T. vulgaris*, and less often with *T. helveticus*. *Triturus cristatus* can take larger prey than *T. vulgaris* (Avery, 1968) and the former species also consumes mainly benthic invertebrates while the latter preys more heavily on planktonic ones (Dolmen and Koksvik, 1983). *Triturus cristatus* is more selective than *T. vulgaris* in its use of microhabitats within the pond, although these preferences change seasonally. Overall, *T. cristatus* exhibits a narrower niche breadth than *T. vulgaris* and several niche dimensions are partitioned to varying degrees, including time in the breeding ponds, activity periods (diurnal versus nocturnal) while in the ponds, diet and microhabitat selection (Dolmen, 1988). Feeding niche overlap is lower than that of spatial niche, and resource partitioning is in general greater than that seen between the two smaller species (Griffiths and Mylotte, 1987).

Although it is difficult and perhaps dangerous to generalize about niche partitioning in adult amphibian communities, there is a widely held view that space is usually the most important factor (Toft, 1985). This is exemplified by a study of a community in central Spain in which 10 species of anurans and urodeles could be allocated to three distinct guilds (aquatic anurans; primarily terrestrial species; newts), and in which the spatial dimension was clearly the main factor determining species segregation (Lizana, Perez-Mellado and Ciudad, 1990). Space was also the most segregated niche dimension in a broader study including two anurans and two lizards on a heath in the Netherlands, whereas strong overlaps were seen in the dimensions of food and activity times among these four vertebrates (Strijbosch, 1992).

6.3.2 Communities of amphibian larvae

European *Triturus* guilds have been studied at the larval as well as at the adult stages. As a general rule the more similar the species, the more similar are the larval diets; competition for food in these larval newt guilds is probably rare, however, since they normally inhabit ponds in which food supply is not thought to limit growth rates (Kuzmin, 1991). There are nevertheless some dietary differences, which are probably secondary consequences of different microhabitat selection in the case of the more divergent species; thus, in curious contrast with adults, the larvae of *T. cristatus* are pelagic for most of their lives and primarily feed on planktonic cladocerans, whereas larvae of *T. vulgaris* are mainly benthic and predate invertebrates of the pond floor (Dolmen and Koksvik, 1983). Extensive observational studies of spatial niche partitioning have also been made on North American urodele larvae, especially those of *Ambystoma* species and *Notophthalmus viridiscens* (summarized in Hairston, 1987). *Ambystoma* larvae migrate vertically within the water column at various times during the diurnal cycle, and different species tend to select different microhabitats within ponds. Unlike the situation with European newts, inverse relationships between density and growth rate of *Ambystoma* larvae have been demonstrated in some ponds, implying that competition could be an important structuring force (Stenhouse, 1985b). However, there is also evidence that interspecific predation can be important. Adult *Ambystoma opacum* actively predate eggs and larvae of *A. texanum* and *A. jeffersonianum*, for example, and in natural ponds the larvae of *A. tigrinum* regularly predate those of other ambystomids and of *Notophthalmus viridiscens*.

Complex amphibian communities of up to 10 species occur in parts of southern Spain, and in this situation resources can be partitioned in several ways. Overlap in food niche between the various anuran tadpoles was high irrespective of whether they were primarily benthic or pelagic, but there were differences, albeit minor, in the use of zones within the pond (ranging from the margin to deep, open water) by the different tadpole species. Much more marked was temporal segregation, with species breeding at different times over the period of eight months during which water is normally available. A higher level of discrimination, this time spatial, was also important; substantial differences between species were manifest in their choice of ponds, some opting for permanent water bodies, some for temporary and others for very temporary ones (Diaz-Paniagua, 1985, 1987, 1988, 1990). Similar results have been found with tropical anuran communities: spatial choices between ponds and temporal differences within ponds were both important at a site in Thailand (Heyer, 1973).

Anuran larvae are generally indiscriminate feeders and this resource is unlikely to be partitioned in most communities (Heyer, 1976), but tadpoles exploit resource-rich environments to achieve rapid growth (Wassersug, 1975) and in at least some circumstances are certainly vulnerable to both intraspecific and interspecific competition for food. However, simple observation also suggests that predation may be an important factor (a glance in almost any pond containing tadpoles will reveal a few in the process of being eaten); specialization evidently helps some species (for instance, those with skin toxins to survive in the presence of vertebrate predators) and environmental stochasticity operates with a vengeance on temporary ponds. This highlights the central difficulty with observational studies, which is that although they can quantify niche widths and overlaps quite effectively, they rarely provide convincing evidence about the relative importance of the structuring forces.

6.4 EXPERIMENTAL STUDIES OF COMMUNITY ECOLOGY

6.4.1 Terrestrial and streamside salamanders

Studies on primarily terrestrial urodeles inhabiting the Appalachian mountains of the eastern United States began in earnest in the 1940s and have become a *pièce de resistance* with respect not just to herpetology but to ecology in general. Their main importance lies in the pioneering use of experimental methods in community ecology, and the central features of the work are worth recounting in some detail. Naturally they began with observational studies, and concentrated on various guilds including two large species of terrestrial salamanders (*Plethodon glutinosus* and *P. jordani*), two small species (*P. cinereus* and *P. shenandoah*), and four species of streamside salamanders

(*Desmognathus quadramaculatus*, *D. monticola*, *D. ochrophaeus* and *D. aeneus*) of widely varying sizes. These studies are described more thoroughly by their initiator (Hairston, 1987).

Plethodon glutinosus and *P. jordani* segregate naturally according to altitude, with the latter species inhabiting the higher mountainous regions and with only small areas of overlap between the two. Plots in the overlap regions sufficient to maintain about 30–40 salamanders at natural densities were fenced, and either left as controls or cleared as far as possible of one or other species of indigenous salamander; the responses of the remaining species were monitored over several successive seasons. Removal of *P. jordani* resulted in increased numbers of *P. glutinosus* relative to controls (Figure 6.1), whereas removal of *P. glutinosus* improved reproductive success of *P. jordani* but did not affect its overall numbers. These results indicated that interspecific competition was occurring, and further studies attempted to establish which resource was the subject of this competition. It turned out to be neither food nor foraging microhabitat, but some spatial element which may be nesting sites.

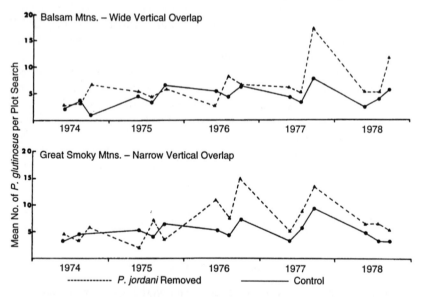

Figure 6.1 Effects of removing *Plethodon jordani* on numbers of *P. glutinosus* in experimental plots at two sites in the USA. (Reproduced, with permission, from Hairston (1980), *Ecology* **61**, 817–826.)

The two small plethodons *P. cinereus* and *P. shenandoah* also inhabit mutually exclusive ranges but in this case the distinction is based on substrate quality. *Plethodon cinereus* is widely distributed but avoids areas of shallow soil such as those found on talus slopes where the much more restricted *P.*

shenandoah lives. Using an experimental approach similar to that of Hairston, Jaeger (1971a, 1971b) showed that *P. cinereus* was a superior competitor on normal, deep soils but that *P. shenandoah* had an advantage on the thinner talus substrates because of its greater ability to retain water. Competition on deep, moist soils is mediated by fighting for refugia, a battle normally won by the more aggressive *P. cinereus*. Fights between these salamanders can be serious; bites are directed to the tail, which often causes autotomy and the loss of fat reserves, or to the snout, which impairs the sensory apparatus and reduces food acquisition (Jaeger, 1981).

Although streamside salamanders often have aquatic larvae, the most detailed studies of their community ecology have concentrated on the adult, primarily terrestrial phase of life. The four species of *Desmognathus* studied by Hairston (1987) and others vary not only in size, but also in microhabitat selection. *Desmognathus quadramaculatus* is the largest and most aquatic; *D. monticola* is smaller and also quite aquatic but prefers smaller streams; *D. ochrophaeus* is smaller still and relatively terrestrial; and *D. aeneus*, the smallest of all, is also the most terrestrial. This pattern was puzzling because it is the opposite of what might be expected on the basis of physiology: smaller animals, with greater ratios of surface area to volume, are normally at the greatest risk of desiccation and might be expected to seek the wettest habitats. Experiments with these salamanders used streamside plots in which, as with *Plethodon*, different species from the guild were removed and the consequences for the others monitored over several seasons relative to unmanipulated controls. The results implicated predation as an important structuring force in these communities. Removal of *D. ochrophaeus* (a potential prey) resulted in a reduction in the numbers of both its larger congeners; removal of *D. monticola*, however, permitted increases of both *D. ochrophaeus* and *D. quadramaculata*, suggesting that the largest two species were in competition, while *D. aeneus* was apparently unaffected by any manipulation of the other three species. Aggression, not only in the form of predation but also for protection of refugia, once again seems to be the mechanism underlying these observations.

6.4.2 Aquatic urodeles

The ease of working with tadpoles has proved a major attraction of amphibians for community ecology and most experiments on aquatic-breeding species have concentrated on the larval phase because growth rates, survival and mass at metamorphosis are all considered to be useful fitness indicators and are simple to measure. Experiments with different densities and mixtures of North American *Ambystoma maculatum*, *A. laterale*, *A. tremblayi*, *A. tigrinum* and *A. texanum* larvae demonstrated complex interactions in which competition and predation both played significant roles. Asymmetry was commonly seen; thus the density of *A. tremblai* larvae affected final weights of *A. laterale* at meta-morphosis, but not vice versa, although each affected the other's development time. Intraspecific interactions were as intense as interspecific ones between three sympatric species (*A. laterale*, *A. tremblayi* and *A. maculatum*) but *A. texanum*, which is not sympatric, was competitively inferior to the other three and is probably excluded from maculatum-group communities by competition intensity (Wilbur, 1972). The larvae of *A. tigrinum* predated those of all three in the sympatric maculatum group, though when alternative food (anuran tad-poles) was available the surviving maculatum group larvae metamorphosed at larger sizes than controls, so predation benefited them by reducing competition.

A simpler two-species system, with *A. opacum* and *A. maculatum*, has also been investigated in some depth. *Ambystoma opacum* eggs develop through the winter months, and on hatching in spring the larvae have a substantial size advantage over those of the sympatric *A. maculatum*. Predation of the latter by the former is very extensive, although it varies between ponds and between years. Field and laboratory experiments showed that predation rates were reduced by the availability of refugia (aquatic vegetation) and the presence of alternative prey, and increased as a function of predator density (Stenhouse, Hairston and Cobey, 1983). Despite the dominance of predation in this sys-tem, *A. opacum* larvae reared with those of *A. maculatum* ended up smaller than *A. opacum* larvae reared alone, indicating some degree of competition even though most of the *A. maculatum* larvae got eaten. In a three-species sys-tem which included *A. opacum*, *A. maculatum* and *A. jeffersonianum*, *A. mac-ulatum* did not respond to the presence of its congeners by changed growth patterns but the developmental period of the other prey species, *A. jeffersoni-anum*, became shorter in the presence of *A. opacum* (Cartwright, 1988).

Evidently there is much variation between ambystomids in their responses to potentially competitive situations which is still far from completely under-stood. Furthermore, studies of larvae alone will probably not be adequate to explain the structures of pond-breeding salamander communities. Interference competition, mediated by aggression, has been demonstrated between adult *A. maculatum* and *A. talpoideum* and may underlie the competitive superiority of the former species in terrestrial habitats (Walls, 1990).

By contrast, rather little experimental work has yet been carried out with *Triturus* guilds. Larvae of *T. vulgaris* and *T. helveticus* suffered slight growth inhibition when raised under food-limited conditions in the presence of the larger *T. cristatus* larvae, but there were no reciprocal effects on the latter species and the dominant effect after four weeks of development, when the size advantage of *T. cristatus* larvae was clear, was predation of the two smaller species (Griffiths, de Wijer and May, 1994). Food is probably not usu-ally limiting in natural ponds, and although ambystomid larvae do have detectable impacts on pond trophic webs (Holomuzki, Collins and Brunkow, 1994) the 'top down' effects of urodeles are in general much less than those of fish. As with ambystomids, the key factor with European newts may be the size relationships, and thus predator–prey interactions, between the species.

6.4.3 Pond-breeding anurans

Anuran tadpoles are particularly convenient organisms for experimental work in community ecology and a great many studies have been carried out with them. Especially important contributions have been made by those using various North American species of *Rana*, *Bufo* and *Hyla*, in submerged cages and artificial replicated ponds.

As a general rule, tadpoles compete for common food resources; at the moderate or high tadpole densities frequently seen in nature, food availability probably limits growth rates in most ponds. Competition between anuran larvae is therefore commonly seen, but is often complex and variable between species. It is usually non-linear, in that doubling tadpole densities does not necessarily double any effects on growth rates or sizes at metamorphosis, and it is frequently asymmetric with very different consequences for the two or more species concerned (e.g. Wilbur, 1982). In some situations, however, tadpoles of species which have sympatric ranges but segregate spatially in different habitats turn out to be competitive equals, as with wood and leopard frogs in North America, inferring that competition during the larval phase is not a prime determinant of niche partitioning (DeBenedictis, 1974). Competitively disadvantaged tadpoles tend to grow more slowly with high rates of variation within cohorts, and normally take longer to metamorphose than superior ones in the same pond; they usually also metamorphose at smaller sizes than siblings not subject to competition. Growth retardation in turn increases mortality risks both from predation, which tends to be inversely proportional to tadpole size, and from pond desiccation.

One particularly important factor affecting the outcome of competition is the timing of reproduction; thus *Bufo americanus* tadpoles in mixed communities grew better if added to experimental ponds before those of *Rana sphenocephala*, while those of *Rana* fared better when added after the *Bufo* larvae (Alford and Wilbur, 1985); when tadpoles of a third, later-breeding species (*Hyla chrysocelis*) were added to the experimental system, they suffered growth inhibition in the presence of *Rana* or *Bufo*, and grew less well than controls even in ponds which had supported *Bufo* earlier in the spring but from which all the toads had already metamorphosed (Wilbur and Alford, 1985). This was interpreted as an impact of the *Bufo* tadpoles on resource availability in the ponds, which especially in ephemeral ones generally declines through the spring and summer season. Similarly, early arrival increased the competitive effects of *Bufo woodhousii* larvae on those of *Hyla crucifer* (in the wild, the hylid normally breeds with or before the bufonid) and the *Bufo* tadpoles always maintained higher activity levels than those of *Hyla* (Lawler and Morin, 1993).

One response to the presence of heterospecific tadpoles is an altered pattern of microhabitat use, such that spatial overlap, and thus presumably competition, is minimized (e.g. Waringer-Loschenkohl, 1988). Spatial considerations may be crucial to the outcome of competition, as shown in experiments with *Bufo americanus* and *Rana clamitans* in artificial ponds (Pearman, 1993). Both species showed lower growth and survival rates in large ponds compared with small ones, and in general competition was more severe in large ponds as a function of increased interior : edge habitat ratios. A third factor of significance can be the water chemistry of the breeding ponds; thus *Hyla gratiosa* and *H. femoralis* show mutual competitive effects, mostly on the length of larval period in the former species and size at metamorphosis in the latter. *Hyla femoralis* was markedly inferior to *H. gratiosa* at pH 6, but performed relatively much better at pH 4.5; in the wild, *H. femoralis* breeds in ponds down to pH 3.5 but *H. gratiosa* only occurs at pH above 4.6 (Warner, Travis and Dunson, 1993). In this case, therefore, specialization along a pH gradient may have operated to minimize competition between two tree frogs with very similar distributions in the south-eastern United States.

Competition between tadpoles may attain special significance when the distributions of normally allopatric species are in a state of flux. The invasion of erstwhile habitat of the natterjack toad *Bufo calamita* by the more widespread and competitively superior common toad *Bufo bufo* and frog *Rana temporaria* is a case in point. Because of changed land management practices, especially a reduction in grazing pressure by domestic livestock, common toads and frogs have encroached upon many sand dune and heathland ecosystems in recent decades and spawned in ponds which were previously the exclusive domains of natterjacks (Beebee, 1977a). Tadpoles of the invasive species cause slower growth (Figure 6.2), smaller size at metamorphosis and increased mortality of natterjack tadpoles at densities comparable with those commonly observed in the field (Banks and Beebee, 1987a; Griffiths, 1991). The invasive species breed earlier in the spring than natterjacks, leading to a size differential among tadpoles which exacerbates the competitive advantage of the invaders, although surviving natterjack tadpoles experience rapid growth retrieval when the other anurans metamorphose and leave the ponds (Griffiths, Edgar and Wong, 1991). Natterjacks avoid spawning in pools already occupied by large numbers of tadpoles, whether these be heterospecific or conspecific, when a choice is available (Banks and Beebee, 1987b) but there is often no such choice once common frogs or toads become well established. The intense competition experienced by natterjack tadpoles in ponds invaded by the more widespread species has probably been a proximal cause of reproductive failure, and ultimately local extinction of natterjacks, at numerous localities in Britain. However, the primary event in this cycle is terrestrial habitat change and the main force operating to partition natterjack toads from common toads and frogs is specialization in the adult phase of the life cycle (Denton and

Beebee, 1994). Clearly it can be misleading to concentrate solely on larval communities in breeding ponds, and at least in this case the primary structuring force operates outside them.

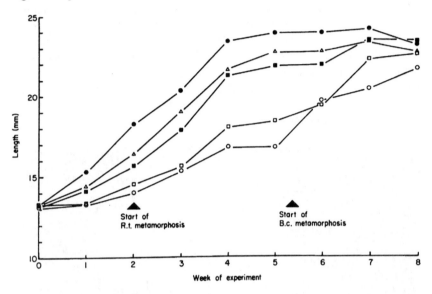

Figure 6.2 Effects of common frog (*Rana temporaria*) tadpoles on the growth of natterjack (*Bufo calamita*) tadpoles in replicate pond experiments. ● = natterjack controls; △ = natterjack tadpoles in water conditioned by frog tadpoles (and including tadpole faeces); ■ = natterjack tadpoles in water to which frog tadpole faeces were added at weekly intervals; □ = natterjack tadpoles in the presence of small frog tadpoles; ○ = natterjack tadpoles in the presence of large frog tadpoles. (Reproduced, with permission, from Griffiths, Edgar and Wong (1991), *Journal of Animal Ecology* **60**, 1065–1076, published by Blackwell Science Ltd.)

6.4.4 Mechanisms of competition between anuran tadpoles

The ways in which tadpoles inhibit each other's growth have been of particular interest since Richards (1958) and Rose (1960) first indicated that interference as well as exploitative mechanisms may be involved. A key feature of the interference mechanism is that direct interactions between tadpoles are not necessary; pond water or sediment 'conditioned' by exposure to one group of tadpoles can cause growth inhibition in a second target group later reared in it, even when the latter are provided with ample food. Alternatively, tadpoles can affect the growth of others separated from them by cage mesh and irrespective of food supply, all implying the production of some kind of water-borne growth inhibitor. Unicellular organisms, replicating in the digestive systems of tadpoles and passed between them via coprophagy (Figure 6.3), have been implicated as mediators of interference-type growth inhibition and although the nature of the organisms has proved controversial, the commonest one seems to be a species of unpigmented algae of the genus *Prototheca* (e.g. Richards, 1962; Steinwascher, 1979; Beebee, 1991; Wong and Beebee, 1994). The action is not species-specific, but when present in tadpole faeces these cells promote coprophagy by small tadpoles and thus divert them away from higher quality food sources (Beebee and Wong, 1992); large tadpoles, though also coprophagous, are apparently not distracted from high quality foods to anything like the same extent.

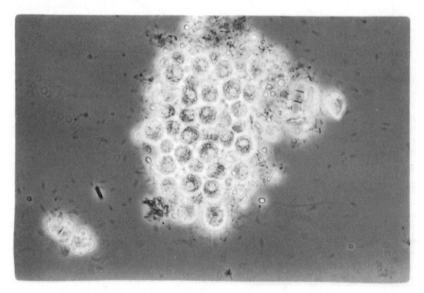

Figure 6.3 *Prototheca*-laden tadpole faeces.

A crucial issue, however, is whether interference competition operates in the field as well as in the laboratory. Evidence on this point is conflicting. Water from ponds in which there was competition between *Rana sylvatica* and *Hyla crucifer* did not inhibit the growth of the competitively inferior hylid tadpoles, inferring direct resource competition as the main mechanism in this case (Morin and Johnson, 1988). Petranka (1989) and Biesterfeldt, Petranka and Sherbondy (1993) looked for growth-inhibitory properties in water from ponds containing high densities of *Rana utricularia* or *R. sylvatica* tadpoles (sometimes in excess of 15 000 per m³), testing it both on tadpoles constrained within the pond and on others in the laboratory. Incidence of interference competition was low on the basis of these tests, with growth inhibition observed in less than 25% of the situations investigated. Moreover, numbers of *Prototheca* cells in tadpole faeces were much lower in wild-caught tadpoles than in those maintained in laboratories. However, studies in replicated ponds with *Rana temporaria* and *Bufo calamita* tadpoles showed clear evidence of interference competition (Griffiths, 1991; Griffiths, Edgar and Wong, 1991; Griffiths, Denton and Wong, 1993) and *Prototheca* counts in the inhibitory faeces were relatively high; moreover, numbers of protothecans comparable with those attained in laboratory experiments have been found in the faeces of *Rana temporaria* and *Bufo bufo* larvae in natural ponds (Wong, Beebee and Griffiths, 1994). On balance it seems that interference competition probably does occur in nature, but relatively rarely and secondary in importance to direct, exploitative competition.

6.4.5 Community ecology of complex systems

Any complete description of community ecology must take into account the interactions of a variety of quite unrelated creatures. Anuran tadpoles, for example, share common food resources with many species of invertebrates in situations where intergroup competition seems likely to occur. Mixed assemblages of aquatic insects, including midge, mosquito and mayfly larvae together with corixids, reduced the mass at metamorphosis of *Hyla andersonii* and *Bufo woodhousei fowleri* tadpoles with effects on the former species being as strong as natural densities of the *Bufo* larvae (Morin, Lawler and Johnson, 1988). By contrast, high densities of *Rana temporaria* tadpoles suppressed growth and egg production in the water snail *Limnaea stagnalis* but the reciprocal effects were positive; snails increased the growth rates of tadpoles, reduced their larval period and increased their mass at metamorphosis, probably by consuming *Cladocera* (an alga not much utilized by tadpoles), thereby increasing nutrient turnover rates and primary productivity (Bronmark, Rundle and Erlandsson, 1991). There is clearly scope for much more work of this kind, which will be essential for a proper understanding of pond community structure.

Rather more studies have been done which include the role of predation, as well as competition, in the structuring of larval amphibian communities. Large predators, especially fish, have major impacts which mainly differentiate on the basis of tadpole unpalatability or toxicity. Of 16 species of North American anurans and urodeles investigated, those coexisting with fish virtually all had some type of chemical defence and those not coexisting generally lacked such protection (Kats, Petranka and Sih, 1988). Curiously, *Ambystoma maculatum* (whose larvae are vulnerable to fish predation) could not distinguish in tests between fish-containing and fish-free ponds (Sexton, Phillips and Routman, 1994) but it remains to be seen whether other species can do so. Less devastating predators, including adult newts, larval urodeles of the larger

species, and invertebrates such as odonate larvae have more complex effects
that can modify the consequences of competition between tadpoles quite dramatically. In a classic study, Morin (1983) showed that in replicated pond
experiments with various combinations of two predators (*Notophthalmus
viridiscens* and *Ambystoma tigrinum*) the outcome of competition between six
species of larval anurans varied substantially according to predation level
(Figure 6.4). In particular, the reduction of tadpole numbers by the predators
greatly improved survival to metamorphosis of two competitively inferior
species (*Hyla crucifer* and *H. gratiosa*), whereas the other four more dominant
species (*Scaphiopus holbrooki*, *Rana sphenocephala*, *Bufo terrestris* and *H.
chrysocelis*) generally fared worse when predators were present. In another
experiment, the effects of competition and predation (by *N. viridiscens*) on
four species of anuran tadpoles were assessed in the context of variable pond
desiccation times (Wilbur, 1987). Predation of *S. holbrooki* larvae again
relaxed competition on other species at high tadpole densities, but was indiscriminate in its effects at low densities; competition in the absence of predation slowed growth and increased risk of mortality from desiccation, while
predation, by relaxing the competition, allowed survivors to metamorphose
before the ponds dried up. There were therefore complex interactions (which
varied between the four anurans) between the effects of competition, predation and environmental stochasticity and the combination of these factors dictated survival to metamorphosis. Predation of one amphibian upon another is
also heavily influenced by relative breeding times. Priority effects in arrival at
breeding sites have a big effect on the predation of *Pseudacris triseriata* tadpoles by those of *Ambystoma tigrinum* (Sredl and Collins, 1991); late arrival
of predator species may result in less predation pressure because the prey are
by then relatively large, and the predatory larvae do not have time to gain the
necessary size advantage.

Invertebrate predation of amphibian tadpoles is rampant in many ponds
and must also be considered in any complete description of pond communities. When predation by *N. viridiscens* and the odonate larva *Anax junius* on
two species of anuran larvae (*Rana palustris* and *Bufo americanus*) were compared, the effects of *Anax* on mortality, lengths of larval periods and growth
rates of survivors were qualitatively similar to but stronger than those of the
newt (Wilbur and Fauth, 1990). A trade-off can exist with respect to pond
ephemerality, which when high minimizes predation by invertebrates but
exacerbates the risk of death by premature desiccation. A particularly interesting relationship exists between the sympatric North American bullfrog (*Rana
catesbiana*) and green frog (*R. clamitans*). The former tends to inhabit permanent ponds and has noxious tadpoles, whereas the latter is more abundant in
temporary pools and produces palatable larvae. The tadpoles were competitive equals when grown together without predators, but in the presence of
caged *Anax* larvae the interactions became asymmetric. Both types of tadpole

reduced their activity levels but the ratio of bullfrog : greenfrog activity increased, favouring relatively more rapid growth of the former (Werner, 1991). However, this greater activity rendered bullfrog tadpoles more vulnerable to predation both by *Anax* and by another predator that inhabits temporary ponds, the larvae of *Ambystoma tigrinum*. By contrast the main predator in the permanent ponds was a fish, the bluegill *Lepomis macrochirus*, which avidly consumed green frog and *Ambystoma* tadpoles, leaving those of the bullfrog alone (Werner and McPeek, 1994). The discrimination of these two species between permanent and temporary ponds is therefore explicable on the basis of larval specialization, with defences targeted against the different types of predator typical of the two environments.

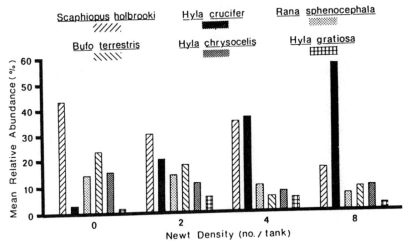

Figure 6.4 Effects of competition and predation on tadpole survival. Results show survival rates of six species of anuran larvae in replicated trials at four different densities of predatory newts *Notophthalmus viridiscens*. (Reproduced, with permission, from Morin (1983), *Ecological Monographs* 53, 119–138.)

6.5 AN OVERVIEW OF AMPHIBIAN COMMUNITY ECOLOGY

The critical dependence of experimental results in community studies on starting conditions, together with the non-linearity of competition and predation effects, highlights the chaotic nature of complex systems particularly in the extreme conditions of temporary ponds (Wilbur, 1990). What generalizations, if any, can therefore be made about the structures of amphibian communities? Specialization has obviously been crucial to the generation of extant species. Where this has led to the occupancy of separate spatial niches, forces such as competition only become relevant in border zones (as with plethodontid salamanders) or during transitions (as with the two bufonids in Britain). A consequence of specialization, however, is often an asymmetry in competitive strength and thus a restriction of the spatial niche breadth of the inferior species below what it might otherwise be; nevertheless, it is important to investigate all life stages of organisms with complex life histories to establish the primary determinants of community structure. Although *Bufo calamita* is competitively inferior to *B. bufo* at the larval stage, specializations of the adults are the primary causes of allopatry between these two species. On the other hand, where several species show strong areas of overlap in the spatial niche, specialization is less relevant to an understanding of their coexistence. Competition will only be important here if a shared niche dimension such as food or space limits the total numbers of individuals within the guild and experience suggests that this may be rather rare under conditions of normal predation rates and environmental stochasticity. Predation, a top-down effect, may be an important structuring force in many amphibian communities where multiple species coexist and further studies of such complex systems are clearly needed.

— 7 —————————————

Distribution, abundance and extinction risk

7.1 DISTRIBUTION

7.1.1 Basic aspects of distribution

A fundamental goal of ecology is to explain the distribution and abundance of species, and this is also of major significance for conservation. In common with most other organisms, frequency histograms of amphibian distributions are strongly skewed with the smallest ones constituting the largest classes. Figure 7.1 shows the patterns of urodele and anuran distributions combining data from North America and Europe, two continents in which the amphibians have been particularly well mapped (Smith, 1978; Arnold, Burton and Ovenden, 1978). Ranges of Eurasian anurans are significantly larger than those of their North American counterparts, with medians of 0.8 million and 0.45 million km² respectively. Urodele range distributions, on the other hand, are similar in both land masses with an overall median of 0.1 million km²; their ranges are on average about five times smaller than those of anurans. Extremes include subterranean urodeles such as the Valdina farms salamander *Eurycea troglodytes*, known from a single sinkhole in Texas, through to the European common toad and frog (*Bufo bufo* and *Rana temporaria*) with global ranges in excess of 5 million km².

The skewed distribution patterns at least partly reflect the abundance of the habitats to which each species is best adapted, and ranges may change dramatically if climatic or other conditions alter. Thus the very restricted high mountain salamanders (*Euproctus* species) of the Pyrenees, Corsica and Sardinia are generally thought to be relics that enjoyed much wider distributions during glacial epochs. Anurans may tend to have larger ranges than urodeles simply because they are more mobile, or it could be for more complex reasons related to different life histories. Anuran larvae are usually omnivorous and those of urodeles carnivorous, a distinction which may permit greater niche breadths for frogs and toads because their larvae exploit abundant primary producers. For both North American and European anurans there are significant positive correlations between range size and body size, although the relationships are weak and explain only a small percentage of the observed variance, and no such relationships hold for urodeles.

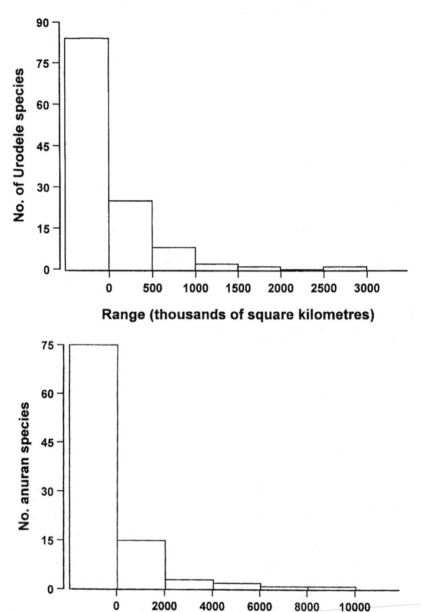

Figure 7.1 Size distributions of anuran and urodele biogeographical ranges in Europe and North America.

It seems clear that the limits of biogeographical distributions are set by climate (though the physiological basis for such limits has rarely been demonstrated in particular organisms) and may often be constrained further by competition or other biological factors. In considering range limits it is necessary to take account of how climate affects the invasion of new sites as well as the maintenance of a species where it is already established, because local extinctions and recolonizations may be frequent in these peripheral zones. Small differences in physiological efficiencies can have large effects on the probability of colonization, and thus may be critical at range edges (Carter and Prince, 1981). The relationship between climate and physiology certainly seems to be important in *Rana lessonae*, one species of amphibian that has been well studied at its northerly outposts (Sjögren Gulve, 1994). However, many factors conspire to create a range border (Hoffmann and Blows, 1994) and correlations may be found with climate, substrate, competitor or predator distributions which can sometimes be tested by physiological studies or transplantation experiments. The salamander *Plethodon cinereus*, for example, avoids acid substrates with a pH of less than 3.8 and this physiological response accounts in part for its distribution in the Appalachian mountains (Wyman and Hawksley-Lescault, 1987). Genetic explanations must also be sought, since the question arises as to why variation has not permitted expansion beyond range borders. Peripheral populations may be in a catch-22 situation, with low variability either because they are small and inbred or because they have been selected directionally in marginal environments, but need high variability to adapt for range extension. Genotypes fit for expansion into new habitats may therefore arise too rarely and, unless well isolated, may in any case be swamped by gene flow from central populations where densities are higher.

7.1.2 Detailed studies of distribution

(a) Obtaining the facts

Distributions are of course ascertained by field survey, a procedure subject to at least two sources of error, notably misidentification and recorder bias. Distribution recording schemes rarely include validation procedures and there is always the risk that distributions reflect the recorders rather than the recorded. On a more positive note, amphibians are relatively easy to locate (especially at breeding time) and their low vagility means that studies are rarely confused by occasional migrants; finding a specimen usually means that a population exists in the immediate vicinity.

The situation in Britain, with its long tradition of natural history, provides a good example. National distribution maps of the six native amphibians have been produced at approximately 10-year intervals since the middle of the twentieth century (e.g. Taylor, 1948; Arnold, 1973; Swan and Oldham,

1993). Coverage improved progressively over this period and the distributions of the British species are currently known over most of the country at 100 km² resolution or better. This historical series should in principle also demonstrate distribution changes that have occurred within the past 50 years, but because of the above caveats such interpretations are only safe where the changes have been large and documented directly, as is the case with the natterjack toad *Bufo calamita* which has declined dramatically in Britain during recent times (Beebee, 1977a). Quite the wrong impression could be gained by uncritical comparisons of the data on common frogs in Figure 7.2, which at face value suggest a substantial increase in abundance over the past 50 years. What the figures really show, of course, is the substantial increase in recording effort during this period.

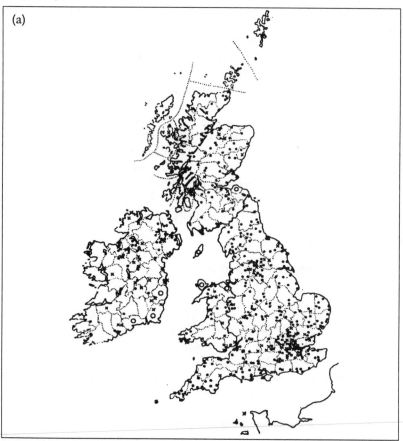

(a)

Figure 7.2 Distribution of the common frog *Rana temporaria* in Britain as known in (a) 1963 and (b) 1995. ● represents positive record. (Reproduced, with permission, from: (a) Taylor (1963), *British Journal of Herpetology* 3, 95–115; (b) Biological Records Centre, Monks Wood.)

National recording schemes are commonly supplemented with more finely scaled studies aimed at generating precise distributions on a local basis or complete distributions of rare species. Examples of the former are legion and in Britain include, among many others, those of Halfpenny (1978) for Staffordshire (4 km² resolution), Buckley (1989, 1991) for Norfolk (4 km² resolution), Langton (1991) for the London area (1 km² resolution), and specialized studies on particular species such as that on *Triturus cristatus* in Huntingdonshire (Cooke, 1983) listing all known breeding ponds. The distribution of the only rare British species, *Bufo calamita*, has also been determined at high resolution such that virtually all breeding sites are probably now known (Banks, Beebee and Cooke, 1994).

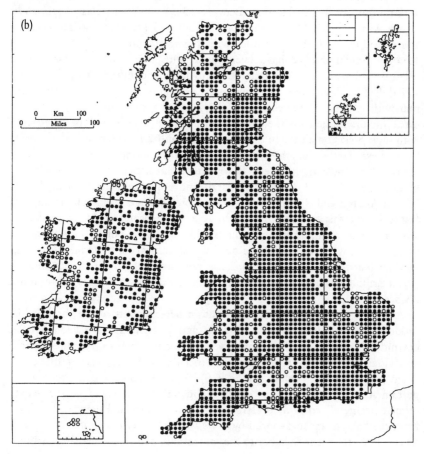

Figure 7.2 *contd* (b)

(b) Interpretation of the facts

Current amphibian distributions are probably explicable on the basis of two separate processes: forces acting at the present time, including climate, habitat structure, competition and predation; and historical events, including colonization routes and local extinctions due to environmental stochasticity.

Attempts have been made at various levels of scale to explain the distributions of the six British amphibians on the basis of these processes. Factors which correlated with the distributions of the five common species over parts of south-east England were identified using discriminant statistics, and indicated that terrestrial habitat characters such as vegetation structure and geology were generally more important than features of breeding ponds in determining local distributions (Beebee, 1985). Although classification efficiency was high by this method, predictability of presence was more erratic and varied from 30–40% accuracy with *T. cristatus* to > 80% for *Rana temporaria*. Similar studies have been made with more complex assemblages (Figure 7.3) such as those that occur in mainland Europe (e.g. Strijbosch, 1979, 1980a; Pavignano, Giacoma and Castellano, 1990) and it seems likely that with greater refinement, particularly in the measurement of habitat features, factors influencing the distributions of many species will be determined with high precision although this is of course only a first step in understanding how these factors operate. The sixth British species, *Bufo calamita*, is restricted in most of northern Europe to small areas of distinctive habitat types, especially sand dunes, heathlands and upper saltmarshes (e.g. Beebee, 1977a; Andren and Nilson, 1985). These biotopes share several physiographical features, and this toads's distribution is evidently constrained by specialized adaptations to open, unshaded habitats with ephemeral ponds. This specialization in turn involves an ability to burrow and thus avoid desiccation, together with an inability to catch prey in dense vegetation; the other British anurans, which are much rarer on these habitats, have the converse traits (Denton and Beebee, 1994).

Attempts to identify critical distribution determinants of the three British urodeles *Triturus vulgaris*, *T. helveticus* and *T. cristatus* have been pursued beyond the resolution of preliminary discriminant analysis. All three species are widespread but *T. helveticus* often occurs in the absence of the others when groundwater is acidic (pH < 6.0) or oligotrophic – conditions which pertain in many mountainous districts and on some lowland heaths. At the other extreme, *T. cristatus* prefers nutrient-rich ponds and tends to be absent from the habitats most favoured by *T. helveticus*. *Triturus cristatus* and *T. vulgaris* frequently occur together in the same pond, but the former species shows an additional preference for pools with extensive areas of open, unvegetated water (Cooke and Frazer, 1976; Yalden, 1986; Denton, 1991). Pond chemistry and physiography therefore seem to be important, but complete explanations of British newt distributions remain elusive. Low pH (4.5), for example, has similar effects on embryos and larvae of *T. vulgaris* and *T. helveticus* (Griffiths, deWijer and Brady, 1993), and although only the latter species seems to breed at low pH in the wild in Britain, *T. vulgaris* breeds in acid pools elsewhere in Europe where *T. helveticus* is absent. Furthermore there are large areas of central and eastern England where *T. helveticus* is absent for reasons which cannot be explained on the basis of current knowledge, and some form of pond chemistry-related competition between these newts remains a distinct possibility. Fish predation of *T. cristatus* larvae may also be a significant factor in local distribution, with cycles of newt and fish (especially stickleback *Gasterosteus aculeatus*) occupation sometimes alternating indefinitely where many ponds are close together (Grayson, 1993). A mixture of terrestrial habitat structure, geology, competition and predation may therefore be important in determining newt distributions in Britain whereas neither climate nor history seem to be significant factors.

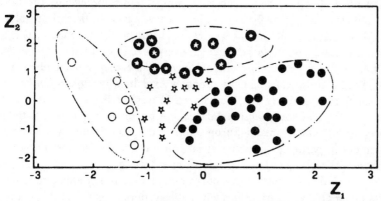

Figure 7.3 Discriminant analysis of *Rana dalmatina* distribution. Canonical discriminant functions (Z_1 and Z_2) are plotted for 61 ponds in north-west Italy. ● = species absent; ◉ = < 50 individuals present; ☆ = 50–100 individuals present; ○ = > 100 individuals present. (Reproduced, with permission, from Pavignano, Giacoma and Castellano (1990), *Amphibia–Reptilia* **11**, 311–324.)

History and climate are, however, probably of critical importance with respect to the distribution of the rare toad *Bufo calamita*. Many areas of apparently suitable habitat exist, particularly in south-west England, Wales and Scotland from which *B. calamita* has never been recorded; some of the more northerly dunes may lie beyond the climatic range tolerable by natterjack toads (which require warmer ponds than the other species for successful tadpole development), but those in the south-west certainly do not. Herpetofauna distributions can become established over evolutionary and ecological time as a result of at least three processes: the first and oldest is the splitting of ancient landmasses, such as Gondwanaland, maintaining original endemics on the fragments thus formed; the second is colonization by various methods of natural dispersal; and the third is artificial introduction by humans. All three have been invoked to explain herpetofauna distributions on islands of the south-west Pacific region (Bauer, 1988), but in Europe the various glaciations and interglacials of the Quaternary period have generated ebbs and flows of amphibians primarily by natural dispersal (Holman, 1993). During the interglacials, amphibians moved north or north-westwards from Ice Age refugia in southern Europe or Asia, with some species crossing land-bridges to Britain and Scandinavia before sea levels rose sufficiently to isolate them. Present herpetofauna distributions in Europe were probably established during the rapid climatic amelioration at the start of the Flandrian period, when the glaciers retreated some 10 000 years BP (e.g. Yalden, 1980; Holman and Stuart, 1991).

Chance has therefore had a substantial impact on these events; the most rapid colonists made it to Britain before sea levels rose sufficiently to block further immigration, but several species extant in northern France (such as *Alytes obstetricans* and *Hyla arborea*) survive perfectly well when introduced to Britain and are presumably absent only because they moved north too slowly. Similar historical accidents probably explain why *Bufo calamita* did not reach some apparently suitable habitats in western Britain, but did colonize a small area of south-west Ireland where populations still persist today (Beebee, 1984). On mainland Europe, too, the compositions of amphibian communities in particular localities at present must be due at least in part to the differential successes of colonization routes from glacial refugia, modified by topographical constraints such as high mountain ranges. Deliberate human impacts on amphibian distributions have, at least until recently, probably been negligible by comparison. On the Balearic Islands, however, there is considerable circumstantial evidence (molecular and sub-fossil) that green toads *Bufo viridis*, and possibly other species, were introduced by humans during the Bronze Age perhaps 3000–5000 years BP (Hemmer, Kadel and Kadel, 1981).

There is unfortunately very little information concerning the most recent historical aspects of amphibian distributions, but there may well have been considerable fine-scale changes over relatively short time spans. At one locality excavated by archaeologists in central England, common toads *Bufo bufo* were apparently rare relative to frogs *Rana temporaria* in the eighth and ninth centuries, whereas by the fourteenth century (and through to the twentieth) toads became abundant, possibly following ox-bow lake formation and thus the generation of a good breeding site nearby (Raxworthy, Kjolbye-Biddle and Biddle, 1990). Very short-term land use changes, generating ponds of different ages, can have substantial effects on community compositions in local areas that affect species distributions over the smallest of temporal and spatial scales (Laan and Verboom, 1990).

7.1.3 Distribution: an overview

Amphibian distributions are determined by a mixture of biology and history; they are often in states of flux in both evolutionary and historical time, and thus cannot always be explained by community ecology alone because community theories mostly assume a state of equilibrium in species composition that rarely pertains other than over short timescales. Nevertheless, the forces invoked in community ecology obviously have important roles in the determination of species distributions. Specialization for particular habitat types limits colonization to varying degrees, and species are ultimately confined by various combinations of climate, competition or predation. The role of history is in large part played out by the changing availability of the habitat type to which a species is confined by its specializations. In the past this was mostly determined by global phenomena (plate tectonics, ice ages and so on) outwith biology, but more recently habitat structure over much of the planet has been dictated by the landscaping of human management. Not surprisingly, this change is having considerable impact on the distributions of amphibians together with many other life forms.

7.2 ABUNDANCE

Distribution is one measure of the success of a species; abundance is another. It is possible to have a wide distribution but to be nowhere common, though this is unusual among the amphibia, probably because sparsity and low vagility are not easily reconciled. Nevertheless, populations do vary in size from place to place and it is important to determine why this is so and thus how populations are regulated. Common frogs (*Rana temporaria*) and toads (*Bufo bufo*) in Britain, for example, assemble to breed in colony sizes ranging from fewer than 10 to more than 1000 individuals (Cooke, 1975a).

Complex life cycles, with larval and adult stages usually frequenting different habitats, make the study of population regulation in amphibians more difficult than with other vertebrate groups. Wilbur (1980) suggested that three types of situation may occur:

- Type 1: species living in restricted habitats, or which are short-lived as adults. These should be regulated primarily at the larval stage. In this situation, adults saturate the larval habitat but not vice versa; larval densities in ponds are high, and inversely related to metamorphic success because of intense competition. The adult population varies independently of larval population size.

- Type 2: long-lived species, or those where adults live in habitats less productive than larval ones. In this case adult numbers may remain too low to saturate the larval habitat; larval survival is determined by stochastic events, whereas adult fecundity and survival are related to adult population density.
- Type 3: species in which adult numbers are high enough to saturate larval habitats, and larval habitat productivity is high enough to saturate the adult habitat. In this situation the population may be regulated at both larval and post-metamorphic stages.

Unfortunately, there have been few studies designed to find out how amphibian populations are actually regulated. Data accumulated over seven years on a population of wood frogs *Rana sylvatica* showed that numbers of breeding adults varied 10-fold, and were clearly related to previous juvenile recruitment rates. Metamorphic success was in turn inversely proportional to the numbers of eggs laid, so adult frog numbers were largely determined by density-dependent factors acting on larval survival in a situation considered to approximate most closely to the Type 3 regulation outlined above (Berven, 1990). Long-term studies of natterjack toads *Bufo calamita* at multiple sites in Britain have shown that adult population densities of this species vary by an order of magnitude or more between sites (Denton and Beebee, 1993b). These densities correlate with toadlet production, which in turn correlates with pond abundance, but toadlet production within ponds is usually positively related to spawning effort; density-dependent intraspecific competition seems rare. Tadpole mortality, and thus ultimately adult population density, is apparently a function of stochastic events in the breeding ponds and thus most similar to Type 2 conditions. However, it seems likely that different populations of the same species can be regulated in different ways according to circumstance and the three regulatory situations identified by Wilbur may represent particular points in a continuum. Thus natterjack toadlet production varies according to the quality of the breeding ponds (e.g. Banks and Beebee, 1988; Sinsch, 1989), and the ratio of pond productivity to terrestrial habitat availability probably dictates which type of population regulatory mechanism is most important. At one extreme are sites with a single, small pond surrounded by large expanses of terrestrial habitat; under these conditions adults saturate the larval habitat and something akin to Type 1 regulation probably occurs. At the other extreme are sites with large numbers of productive ponds in moderate areas of terrestrial habitat, in which both become saturated in a Type 3 situation. Animals living at high densities in these rather rare Type 3 sites, unlike those at more typical ones, show very weak correlations between age and body size and many animals remain small with low fecundity throughout life.

7.3 EXTINCTION RISK

7.3.1 General aspects

Despite their obvious significance in the evolution of life, extinction processes have received surprisingly little study (Raup, 1991). It has been estimated that for every species alive today at least 1000 have died out since the origin of multicellular organisms in the pre-Cambrian era, and species are thought, on the basis of the fossil record, to have an average lifespan of about 4 million years. We should therefore expect one amphibian to go extinct every 1000 years if there are, as currently estimated, approximately 4000 extant species. As might intuitively be expected, organisms with small populations or restricted distributions seem much more likely to die out than those with the opposite properties. For widespread species to become extinct some kind of 'double hit' may be necessary, notably a 'first strike' to reduce range and a second blow (often of a quite different character) to finish the job. Extinctions might be precipitated by at least three types of process: completely random ones (Raup's 'field of bullets', striking totally indiscriminately); ones that are selective on the basis of some kind of Darwinian fitness measure; or 'wanton' ones, meaning random with respect to fitness but selective on some other basis not related to how well an organism is adapted to its environment. In the latter situation, which may be the commonest, bad luck is much more important than bad genes. This scenario is highly pertinent to the current threats to biodiversity posed by human activities; there is increasing evidence that the biosphere is entering into a new phase of mass extinctions which may be even larger than presently realized because of time-lag effects from habitat destruction that are not yet apparent (Tilman *et al.*, 1994).

Attempts to ameliorate mass extinctions among amphibians will require consideration of many factors in common with other groups of organisms. Firstly, there is a widely held view that maintaining species is not enough; smaller degrees of genetic variation, such as subspecies, also warrant consideration. A need to identify evolutionary significant units (ESUs) for conservation, possibly based on molecular criteria such as complete monophyly in mitochondrial DNA alleles, has been proposed (Ryder, 1986; Moritz, 1994) but may not be practicable for use on the scale likely to be necessary. More subjective criteria, such as degree of geographical isolation and minor morphological variations, will probably continue to feature strongly in this debate for many years to come. Secondly, since rarity is directly related to extinction risk, it is important to consider the specific dangers inherent in rarity. At least five such dangers, which may act synergistically, can be identified:

1. Deterministic threats directly consequent upon human activities, including habitat destruction, pollution and direct predation. These are considered further in Chapter 8.

2. Environmental stochasticity, such as a series of drought years exterminating amphibian populations dependent upon temporary ponds. The disappearance of some frog species from parts of south-east Brazil may, for example, have been caused by unusually severe frosts in 1979 (Heyer *et al.*, 1988).

3. Demographic stochasticity. The 'gamblers ruin' model (Raup, 1991) predicts that, given long enough, every isolated population will eventually go extinct if not supplemented by occasional immigration. This is because upward oscillations in population numbers can always fall again no matter how high they rise, but oscillations down to zero are irreversible. Studies of isolation effects on Swedish pool frogs (*Rana lessonae*) suggest that practice can follow this theory only too well (Sjögren Gulve, 1994).

4. Loss of genetic diversity. There is a substantial body of theoretical work on this subject, but as yet little to demonstrate its importance to amphibians. Small, isolated populations are bound to lose alleles by genetic drift and in extreme cases are likely to become inbred, all of which can reduce viability of individuals as well as the ability of the population to survive changes in environmental conditions.

5. Living in the wrong place. Habitats maintaining maximum biodiversity, and thus prime candidates on one important criterion for protected status, are often not the places in which rare species live (Prendergast *et al.*, 1993). Species-rich areas do not coincide for different taxa, and there are certainly rare amphibians that live in habitats unlikely to attract attention on any other basis (such as, for example, the subterranean salamanders).

The realization that small populations are at particular risk has led to deliberations about how big they should be to stay out of trouble, and thus the concept of minimal viable population size (MVP), i.e. the number of individuals necessary to give a 95% probability of a population persisting for at least 1000 years. Theoretical approaches to MVP have attempted to combine both demographic and genetic consequences of being small (e.g. Nunney and Campbell, 1993), and indicated that MVPs will normally be in the region of at least hundreds and often thousands of adults. Circumstantial evidence suggests, however, that many isolated amphibian populations must have survived for millennia with much smaller average sizes than these. There may therefore be theoretical considerations not yet fully taken into account in MVP calculations for amphibians, but in any case it seems certain that an ability to persist in small numbers will be ever more important in future.

Although rarity and small population size have received most attention by conservationists, a second indicator of difficulties ahead is a rapid decline of taxa still relatively abundant. These organisms could be undergoing a 'first strike' and prudence dictates that efforts should also be directed towards species in this condition. Paradoxically, there are more practical studies of this situation (in which underlying theory is poorly developed) than there are of problems with small populations (Caughley, 1994), and there are many examples of such declines among the amphibia. Unfortunately it usually costs more to arrest declines of widespread species than it does to protect a few sites with very rare ones, and it can be difficult to instil the same sense of urgency into decision-makers.

7.3.2 Risks to European amphibians

The situation with respect to European amphibians exemplifies the difficulties of assessing extinction risks. The first attempts to compile a list of threatened European amphibians were made in the 1970s and resulted in the Berne Convention of 1979, in which those species considered most endangered were listed on Appendix II (Appendix I dealt with flora). No attempt was made to distinguish cases for conservation below the species level, and five species of urodeles (about 20% of the total recognized at that time) were placed in Appendix II. The median range of these species, at 80 000 km², was lower than that of all European urodeles considered together (140 000 km²); thus there was some selection on the basis of small range size though one of the five species, *Triturus cristatus* (which then included the subsequently differentiated *T. dobrogicus*, *T. carnifex* and *T. karelini*), had the largest global range of all the European urodeles. Also, just over 50% of anuran species recognized in 1979 were placed on Appendix II and in this case the median range size of the protected species was not significantly different from (but actually slightly larger than) that of the entire European anuran fauna taken together. Many of the amphibian designations were made on the basis of very limited information about a group which, at that time, was still very much under-recorded. The data came exclusively from countries constituting the Council of Europe, which are mostly the westerly ones, and thus took little or no account of areas which for some species comprised the majority of their global ranges. The allocation of *Triturus cristatus* to Appendix II, for example, was largely based on observations covering well under 20% of the newt's total distribution. Of course widespread species may be undergoing a 'first strike' and it will always be important to consider them as well as those with inherently small ranges, but even so it is difficult to understand how placing more than 50% of anurans on Appendix II was justifiable.

A more scientific attempt to assess the status of European amphibians began almost immediately after the Berne Convention with the work of the Societas Europaea Herpetologica (SEH). This included not only compilation of existing and historical distribution data but also extensive fieldwork by a team of professional herpetologists and, crucially, taking into account east as well as west European countries (though not the old Soviet Union) as far as practicalities permitted. Subspecies as well as full species were considered and the results of this work, with new lists of priority endangered species, were published by Corbett (1989).

Although details of the analysis are not given, it is clear that both smallness of range and perceived rate of decline were used to produce lists much better differentiated than those of the Berne Convention. Category 1, considered to be the most endangered taxa, included just two urodeles and three anurans (excluding a Turkish species). All five taxa survived only in small home ranges (less than 70 000 km^2) and at least three were known to be in decline; two of the five (the pale alpine salamander *Salamandra atra aurorae* and the Italian spadefoot toad *Pelobates fuscus insubricus*) were recognized as subspecies rather than full species at the time of listing. Several species with similarly small ranges, such as the three mountain salamanders (genus *Euproctus*), the plethodontids (genus *Speleomantes*) and some of the painted frogs (genus *Discoglossus*) were classified as vulnerable or rare. Only one species in this revision, the fire-bellied toad *Bombina bombina* which is also listed as vulnerable or rare, appears markedly anomalous. The global range of this anuran is more than 1.3 million km^2, and the reason for its inclusion seems to be based on a decline within Council of Europe countries (for which it is listed as one of only four key amphibian species), despite the fact that almost all its range lies further east. This toad, with the closely-related *B. variegata*, and *Triturus cristatus* also feature on Annex II of a European Economic Community Directive of 1992 in which all three are given maximum protection together with the much more localized species of amphibians listed as endangered or rare by Corbett (1989).

Rarity or decline assessments based on arbitrary political boundaries are surely dubious by comparison with those based on global status, and the kinds of problems faced by relatively widespread species might be better addressed at a more local (i.e. nation state) scale, as was also done by Corbett (1989). Virtually all species become rare, and are therefore bound to suffer occasional local extinctions, near their range edges; it is important not to bias assessments of overall rarity with events at the extremities. Failure to distinguish between global and local rarity (or widespread slow decline) risks devaluing efforts to save species in the former, truly perilous category. On the other hand, identifying what might be an ominous first-strike type of decline in a widespread species is also important but inherently very difficult. Inevitably it requires study over large geographical areas, often for long periods, to provide convincing evidence and this has yet to be done properly for any European amphibian.

Essentially similar approaches to those described for Europe have been adopted elsewhere, including in the United States where 5–13 amphibians recognized as the most endangered by three separate assessment agencies all have relatively small geographical ranges (Bury, Dodd and Fellers, 1980). What matters, of course, is whether recognition leads to salvation.

—— 8 ————————————————————

Threats to amphibians

8.1 BACKGROUND

Ultimately the main threat to the future of amphibians today is the same one that faces other life forms on the planet: the continued increase in numbers of one species, *Homo sapiens*, together with its extraordinarily high use of natural resources in the most developed parts of the world. The tackling (or otherwise) of these issues will ultimately decide the fate of all biodiversity, but in the short term it is crucial to identify those aspects that affect amphibians most directly. This is not always easy, as the fate of the Wyoming toad *Bufo hemiophrys baxteri* illustrates. Formerly widespread in its namesake state, it declined rapidly during the 1970s to a single small population by the early 1980s. Postulated reasons for this plunge towards extinction include insecticide sprays, herbicide applications, changed land-use (especially drainage methods), increased predation by gulls and foxes, and disease; but nobody really knows (Baxter, Stromberg and Dodd, 1982; Beiswinger, 1986). It is remarkable that until very recently amphibians seem to have survived well and between 1600 and the 1970s there is evidence of only two extinctions, compared with 28 for reptiles (Honegger, 1981). However, recent concern about global amphibian declines (Chapter 2) are changing this picture rapidly and it is ever more pressing to evaluate deterministic causes of declines and local extinctions. At least four main threats can be identified, and these are discussed in decreasing order of importance.

8.2 HABITAT DESTRUCTION OR ALTERATION

The distinction between loss and alteration of habitat is not always an easy one to make. Destruction usually entails dramatic changes such as replacing green fields with housing estates, but converting one type of green field into another (such as pasture into arable) can be just as terminal for many species. Irrespective of the niceties of definition, there can be little doubt that habitat change has been of primary importance in most amphibian declines throughout the world. This is presumably true in many tropical regions where amphibian species diversity is at its highest but where forest clearance continues at a terrifying pace; unfortunately there has as yet been little attempt to evaluate the effects of these clearances on wildlife. However, logging operations in the temperate woods of North America clearly threaten both the rare larch mountain salamander *Plethodon larselli* in Washington state, and the Red Hills salamander *Phaeognathus hubrichti* in Alabama, probably by reducing soil moisture levels below tolerable levels (Aubry, Senger and Crawford, 1987; Dodd, 1991).

Damage to amphibian populations has certainly been substantial in some of the less pristine rural habitats of the developed world. Drainage of marshland to improve crop yields and to reduce diseases such as liver fluke and malaria has continued apace, especially since the seventeenth century, to the extent that in Britain (for example) only isolated patches now remain (Rowell, 1986) and 82% of marshland in east Essex was destroyed between 1938 and 1981 alone (Williams and Hall, 1987). Field ponds, on the other hand, experienced high popularity as watering places for livestock until the sharp swing in favour of arable farming that began about the middle of the twentieth century. Since that time the losses of this favourite amphibian breeding habitat have been dramatic: in the 1880s it is estimated that Britain was home to some 1.3 million ponds, but this was reduced by more than 70%, at an average loss rate of 9000 per year, to perhaps 375 000 ponds by the 1980s (Oldham and Swan, 1993). In the face of this, it is scarcely surprising that amphibians declined and in one severely affected part of Britain frogs and toads were reduced by the 1970s to perhaps 1–2% of their levels in the 1930s (Prestt, Cooke and Corbett, 1974). Although a substantial number of ponds have been deliberately infilled, many more are lost by the process of natural succession; it requires positive management to sustain the existence of a pond, and the economic imperative for this has now largely disappeared in most developed countries.

On top of that, changes in farming practice have also had severe adverse effects on amphibian terrestrial habitats. Intensive management of arable land (Figure 8.1) now involves not only regular ploughing but also pesticide and fertilizer applications, as well as highly mechanized reaping at the end of each cycle. This creates an inhospitable environment for adult and juvenile amphibians during their terrestrial life phases, whether foraging in summer or hibernating in winter. Of course some animals do survive on arable land, but usually at much lower numbers than ranker, less disturbed habitats are able to maintain (e.g. Beebee, 1977b; Verrell, 1987b). Satellite-based monitoring methods are beginning to produce quantitative data on the rate and extent of habitat change, including alterations to farmland, that should in future permit more thorough assessment of this most serious type of damage. To cap it all, increased habitat fragmentation by roads and railways can effectively isolate surviving amphibian populations and increase the risk of genetic deterioration (Reh and Seitz, 1990).

Not all deleterious changes are as obvious as pond loss or arable desertification. More subtle, but potentially just as dangerous, can be the accidental or deliberate spread of various animal or plant species. Thus the introduction of predatory fish, usually for angling purposes, has rendered many ponds useless to amphibians especially sensitive to fish predation (such as *Rana cascadae* in North America, or *Triturus cristatus* in Europe) and caused innumerable local extinctions as a result. There are also instances of amphibian translocations known or suspected to have damaging effects on native species, including the establishment of North American bullfrogs *Rana catesbiana* in several parts of Europe, the spread of marine toads *Bufo marinus* from central America to many other parts of the tropics, and the widespread escapes of clawed toads *Xenopus laevis* that have formed feral populations in Europe and North America. These animals are large predators which, when they become abundant, constitute a habitat change that could have serious impacts on various elements of the native fauna. Alien plants, too, are altering habitats on an ever increasing scale. Commercial plantations of Australian eucalyptus in Portugal have reduced the species-richness of mammals, birds, reptiles and plants but apparently have not significantly affected amphibians, which is just as well since the restricted golden-striped salamander *Chioglossa lusitanica* is endemic to the area concerned (Vences, 1993).

Figure 8.1 Intensive arable farming destroys amphibian terrestrial habitats. (Photo: T. Beebee.)

Amphibians inhabiting unusual or uncommon habitats are even more at risk from changes than are more catholic species, because such places are often peculiarly sensitive to external influences. The New Jersey Pine Barrens, for example, have suffered wetland degradation and an influx of alien plant species as a result of groundwater eutrophication from a relatively small increase in the native human population (Ehrenfeld, 1983). In South Africa, the rare *Xenopus gilli* has been lost from 60% of its recent localities within the past 50 years because of disturbance to and loss of its unique blackwater (peat-stained) breeding ponds, with consequent encroachment of, and destructive hybridization with, the abundant *X. laevis* (Picker and deVilliers, 1989). Sand dunes, with their ephemeral ponds, constitute important amphibian habitats all over the world but many have suffered extensive degradation from scrub invasion, water-table lowering and eutrophication (e.g. Fuller and Boorman, 1977; van der Meulen, 1982). Scrub species such as sea-buckthorn (*Hippophae rhamnoides*) and rhododendron (*Rhododendrum ponticum*) have spread, following deliberate introductions, to transform sand dune and heath-land habitats from open to overgrown environments in many countries. This has been to the detriment of native animals and plants, including amphibians such as the natterjack toad *Bufo calamita*. The irony is that the spread of alien scrub in Britain was promoted by decimation of the rabbit (*Oryctolagus cuniculus*), another introduced species, by an introduced myxomatosis virus. Events of this kind demonstrate the awe-inspiring complexity of events trig-gered by human intervention in recent decades.

Heathlands, particularly those of northern Europe, are another habitat type in serious trouble. Their expanse was reduced in Denmark by about 85%, from an estimated 1.1 million ha in 1850 to perhaps 0.17 million ha in 1966, primarily by reclamation for agriculture (Joensen, 1967). In Dorset, once a heathland stronghold in Britain, the area of this habitat declined by more than 85% (Figure 8.2) from some 40 000 ha in 1750 to about 5000 ha in 1987 (Webb and Haskins, 1980; Webb, 1990). These losses were due to recla-mation for agriculture, afforestation with conifers, urbanization and scrub encroachment and have been so severe that the specialist *Bufo calamita* was almost eliminated from this habitat, half its historical range in Britain, during the past 100 years (Beebee, 1977a). Natural succession has become the great-est threat to heathlands (e.g. Marrs, Hicks and Fuller, 1986), and it seems likely that only a return to historical management practices, especially low-density livestock grazing, can save what remains of these habitats and their fascinating biota.

Not all habitat change spells doom for amphibians. Old mine workings can, when abandoned, develop into excellent amphibian habitats especially when pools develop on the quarry floors. In Australia, disused bauxite mines are rapidly reclaimed by eucalypt forest and within 4–6 years may support as many frog species as prior to the disturbance (Nichols and Bamford, 1985), while European yellow bellied toads *Bombina variegata* thrive in old gravel pits (Barandun, 1990). Similar stories could be told for many species in many countries. Sadly, however, sites of this kind are now often under pressure for landfill with domestic rubbish and thus are not by any means a certain reser-voir for amphibian populations in the future.

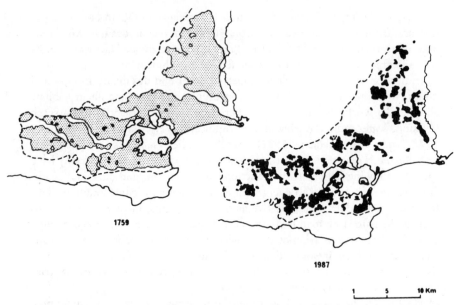

1759

1987

1   5   10 Km

Figure 8.2 Reduction in heathland habitat in Dorset, England, between 1759 and 1987. (Reprinted from *Biological Conservation* 47, N.R. Webb, 'Studies on the invertebrate fauna of fragmented heathland in Dorset, UK, and the implications for conservation', p. 155 (1989) with kind permission from Elsevier Science Ltd, The Boulevard, Langford Lane, Kidlington OX5 1GB, UK.)

Perhaps the most surprising success story has followed the spread of suburbia across the countryside. A regular accompaniment of this apparently destructive development has been a vogue for artificial garden ponds. These have been the veritable saviours of some amphibian species, appearing at a time when changes in farming methods were devastating populations in the wider countryside. In one small part of southern Britain there were estimated to be several thousand garden breeding sites within a few square kilometres, probably more than at any time in recorded history, by the 1970s (Beebee, 1979). There is every reason to believe that other towns and cities are similar in this respect, but of course individual ponds (and thus amphibian populations) tend to be small and not all species adapt to the urban habitat.

8.3 USE OF AGROCHEMICALS

8.3.1 Pesticides

Within a decade of the start of the mid-twentieth century's agricultural revolution came the realization that one of its mainstays, a new generation of chemical insecticides and herbicides, was having devastating effects on wildlife and the legacy of this era of unregulated poisoning is still with us. Amphibians, with their highly permeable skins, could be particularly vulnerable to pesticide sprays and there have been many studies to ascertain whether this expectation is realized. Investigations tend to be of two types: the first and relatively easy kind is to determine, usually by some kind of spectroscopic or chromatographic analysis, whether body tissues contain significant concentrations of the substance in question; the second and much more difficult task is to assess the biological effects of pesticides on an organism, at both the individual and population level.

DDT, an organochlorine insecticide produced in enormous quantities from the 1940s onwards, is the archetypal pesticide and has received more attention than any other. Still in use, though under regulation in most countries, this compound is highly persistent and has been found (often as a partial degradation product, DDE) in tissues of animals living thousands of kilometres from its nearest site of application. It is hardly surprising, therefore, that several species of frog and salamander have been found to contain traces of DDT or DDE. Treatment of common frog *Rana temporaria* tadpoles with acute (relatively high) DDT concentrations leads to developmental abnormalities, hyperactivity and sometimes death; survivors may die later at metamorphosis, probably when DDT residues stored in fat reserves are mobilized (Cooke, 1970). Exposure to lower (chronic) doses for longer periods has much less effect, though it speeds up development so that treated tadpoles metamorphose earlier than controls (Cooke, 1973a). Field trials, in which tadpoles were housed in cages immersed in ditches near to where DDT was being applied, confirmed these results (Cooke, 1973b). Similar data were obtained with North American species such as *Rana pipiens*, but the question remains as to what the overall ecological impact of such treatments might be. Uncoordinated hyperactivity probably makes tadpoles more vulnerable to predation, and such larvae are taken selectively by newts (Cooke, 1971), but of course the pesticide is likely to be having even more serious effects on the main invertebrate predators of tadpoles. Adult amphibians are also affected by high concentrations of DDT, usually becoming hyperactive and again presumably more vulnerable to predators. Cricket frogs *Acris crepitans* in Mississippi have been seen in the field demonstrating the typical symptoms of DDT poisoning (Ferguson and Gilbert, 1967) and it seems very likely that DDT and other insecticides have affected amphibian populations in at least some areas where applications have been heavy.

Herbicides may also pose threats to amphibians. Applications of atrazine, a general-purpose herbicide, to a railway track resulted in contamination of a nearby pond with effects on the common frog *Rana temporaria* population that included massive egg mortality and, at sub-lethal concentrations, tadpole deformities (Hazelwood, 1970). Most herbicides, including for example

Diquat and Dichlobenil, are apparently non-toxic to frog and toad tadpoles but may have indirect effects by promoting algal growth at the expense of macrophytes (Cooke, 1977). Once again the significance of this at the population level is unclear; tadpoles might benefit from the increased supply of algal food, or suffer extra predation from the reduced macrophyte cover.

With the large numbers of pesticides now in use, and the complex indirect effects that can superimpose on direct, readily measurable ones, it will always be difficult to assess the impact of these chemicals on amphibians or any other group of organisms in the field. However, amphibian tadpoles can be useful as bioassays for pesticides since the appearance of developmental abnormalities (Figure 8.3) is a sensitive indicator of sub-lethal effects (Cooke, 1981). There is an ever-increasing database on the susceptibility of amphibians to synthetic toxins (Power *et al.*, 1989) and evidently vigilance will always be necessary in this area. At present, the overview is that amphibians in general seem surprisingly resistant to most pesticides despite the specific dangers discussed above. When tested with a range of organochlorine, organophosphate and carbamate compounds the acute oral toxicity dose for bullfrogs (*Rana catesbiana*) was usually one or two orders of magnitude higher than for mallard ducks (Hall and Henry, 1992). Only two substances used in fish culture have been identified as having especially severe effects on amphibians, notably TFM (3-trifluoromethyl-4-nitrophenol), developed to control sea lampreys, and formalin, specifically used to kill tadpoles in fishponds.

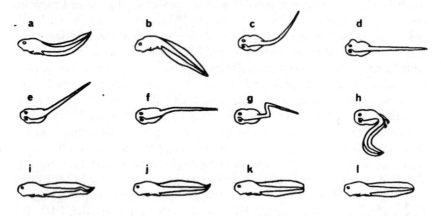

Figure 8.3 Examples of common tadpole deformities, including severe body kinks (e.g. (a)–(c), (h)) and less obvious malformations (e.g. (d), (i)–(k)); (l) is an almost normal tadpole. (Reprinted from *Environmental Pollution* 25, A.S. Cooke, 'Tadpoles as indicators of harmful levels of pollution in the field', p. 125 (1981) with kind permission from Elsevier Science Ltd, The Boulevard, Langford Lane, Kidlington OX5 1GB, UK.)

8.3.2 Fertilizers

Until recently, the effects of fertilizers on amphibians have received much less attention than those of pesticides. Of course the indirect dangers, especially of pond eutrophication from run-off containing high concentrations of nitrates or phosphates, have long been recognized but the observation that high nitrate levels in drinking water can affect human health has prompted more interest in these apparently innocuous compounds.

Although studies are still in their infancy, preliminary evidence suggests that nitrate ions at concentrations regularly attained in natural ponds subject to run-off from fertilized fields (40 mg per litre or thereabouts) are surprisingly toxic to anuran larvae (Baker and Waights, 1993, 1994). Tadpoles of *Bufo bufo* and *Litoria caerulea* suffered significantly reduced growth rates and elevated mortality at this concentration of nitrate ion. Other evidence also suggests that adult frogs are highly susceptible to nitrate poisoning and may often be exposed to nitrate applications that are routinely made at the same time in spring that frogs are migrating to their breeding ponds. These are surprising results with serious implications, and undoubtedly warrant much more study.

8.4 ATMOSPHERIC POLLUTION

8.4.1 Acid rain

As long ago as the 1920s the realization began to dawn, following inexplicable fish kills in Scandinavian rivers, that acid rain and snow had begun to fall. Unpolluted rain, saturated with carbon dioxide, normally has a pH of around 5.6; water contaminated with sulphur and nitrogen oxides generated by the combustion of fossil fuels regularly produces rainfall of pH 3–4. Particulate pollution makes matters worse still by adsorbing on to the leaves of trees, especially conifers, and generating even higher levels of acidity when washed off by subsequent precipitation (e.g. Alcock and Mordon, 1981). Mineral-rich soils buffer this acid load effectively but in impoverished, hard-rock or podsolized landscapes the ponds and rivers present have no such protection. Debate raged for some time as to whether acid rain or land-use changes bore prime responsibility for groundwater acidification, but the issue was largely settled by the development of diatom analysis of lake sediment cores and its demonstration that acidification, often of between 1 and 2 pH units, correlated with industrial pollutant input and not alterations in land-use (e.g. Flower and Battarbee, 1983). Biological damage inflicted by the pH changes typically generated by acid rain include losses of gammarids, failure of fish to breed and the development of filamentous algal mats (Schindler *et al.*, 1985).

Just as with pesticides, many factors conspire to make the study of acidification on amphibians less than straightforward. A critical starting point is knowledge about the pH ranges over which amphibians normally breed, and as might be expected there is considerable variation here between species. In Nova Scotia, for example, three out of 11 species observed in 159 water bodies were found breeding under acid conditions in the field; *Rana sylvatica* and *Ambystoma maculatum* reproduced successfully in acid bogs of pH 4.1, while *Rana clamitans* larvae were seen in pools of pH 3.9 (Dale, Freedman and Kerekes, 1985). Some European amphibians also breed at pHs of around 4, especially the moor frog *Rana arvalis* and palmate newt *Triturus helveticus*, but a further complication is that pH limits in the field often vary for the same species in different parts of its range. Thus the newts *Triturus vulgaris* and *T. cristatus* are very rarely found below pH 6 in Britain, but sometimes breed at pH 4.5 in Norway (Dolmen, 1980). Field observations are only a crude indication of pH tolerance because they are usually not quantitative with respect to egg or larval mortality rates, and often ignore other aspects of water chemistry that might act cooperatively with pH.

Field and laboratory experiments have provided an insight into just how complex the effects of acidification can be. Both low pH and monomeric aluminium (as AlOH or AlF), liberated from organic complexes under acid conditions, contributed to toxicity effects on tadpoles caged in streams and subjected to various chemical treatments (Clark and Hall, 1985). The three amphibian species tested showed differential sensitivities to these two pollutants, with   eggs and embryos generally more susceptible than larvae. Aluminium affects respiration by damaging gill membranes in both amphibians and fish, whereas low pH causes death by generating massive effluxes of sodium ions and, in the case of embryos, by inhibiting the activity of hatching enzymes (Freda and Dunson, 1984, 1985). Sodium efflux can be countered to some extent by high concentrations of external cations such as Ca^{2+}, all of which means that pH effects in the field can be influenced significantly by other aspects of water chemistry.

Interpretation of events in natural ponds is very difficult in the light of these complexities. Pough (1976) anticipated that the future of *Ambystoma maculatum* in the north-eastern United States was bleak because of continuing groundwater acidification, but this prognosis was not confirmed by subsequent study of *Ambystoma* in the Connecticut valley (Cook, 1983). An investigation of 235 water bodies in the Sierra Nevada also indicated that groundwater acidification was unlikely to have been responsible for amphibian declines in that particular area (Bradford *et al.*, 1994) and at present there is little convincing evidence that acid rain has had widespread effects on amphibians in North America. In Europe, most interest has centred on the common frog *Rana temporaria* which frequents, among other habitats, hardrock areas subject to acidification. Its response to acid conditions is typical of many amphibians: fertilization efficiency is little affected down to pH 4, unless inorganic monomeric aluminium concentrations are high (Beattie, Aston and Milner, 1991); eggs and embryos are more sensitive to low pH than tadpoles, the former normally succumbing over the pH range 4–4.5, but at higher pHs (4.5–6.0) in the presence of monomeric Al at concentrations of 50–400 µg per litre (Tyler-Jones, Beattie and Aston, 1989). Decreasing pH below 6.0 progressively reduces tadpole growth rates and maximum body size, and delays metamorphosis, though survival (when protected from predation and desiccation) is high even at pH 3.6; but the presence of monomeric Al at pH 4.4 once again compounds pH effects, also causing developmental abnormalities and significant mortality rates (Cummins, 1986b). Short-term pH depressions, such as can occur following snow melt or episodes of acid rain, are ameliorated to some degree in their effects on *Rana temporaria* embryos by calcium ions at concentrations as low as 4 mg per litre (Cummins, 1988).

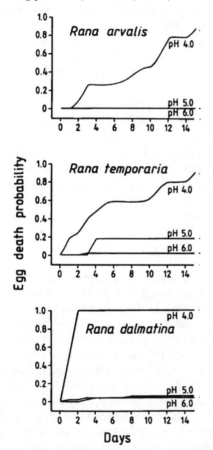

Figure 8.4 Effects of pH on egg mortality of three Swedish anurans. (Reproduced, with permission, from Andren *et al.* (1988), *Holarctic Ecology* **11**, 127–135.)

The important question is: have common frogs suffered from acidification in the field? The decline to extinction of a frog population in a small Swedish lake between 1974 and 1979 occurred in circumstances that looked very like acid pollution (Hagstrom, 1980). High levels of egg and embryo mortality have been seen in a part of south-west Scotland well documented for acidification of its lakes, and where conditions of pH (4.0–4.5) and aluminium concentration (500 µg per litre or above) were sufficient to account for the mortality observed (Cummins, 1986b). Interestingly, though, smooth newts *Triturus vulgaris* apparently increased in the acidified Swedish lake, probably as an indirect consequence of fish being exterminated along with the frogs. Frogs have not disappeared from south-west Scotland, and overall there is no evidence that acidification has caused widespread extinctions of this species; indeed there may even be circumstances in which low pH can improve fitness by reducing density-dependent growth inhibition in tadpoles (Cummins, 1989). Common frogs living in areas with many acid pools are not adapted to reproduction at low pH (Tyler-Jones, Beattie and Aston, 1989) but moor frogs *Rana arvalis*, a morphologically very similar species inherently more tolerant of acidic breeding sites (Figure 8.4), do seem to vary in this tolerance as a function of breeding pond pH (Andren *et al.*, 1988; Andren, Marden and Nilsson, 1989).

A more serious situation arises for species with distributions highly dependent on habitats vulnerable to acidification, such as the European natterjack toad *Bufo calamita* on mineral-poor heathlands. Some heathland ponds in the Netherlands and Britain have acidified in recent decades (Figure 8.5), and in Britain this has correlated with their abandonment as natterjack breeding sites when pH fell below that at which embryos can survive (Beebee *et al.*, 1990). In cases like this, acidification may have had a significant impact upon both distribution and abundance.

8.4.2 Ozone damage

Other consequences of atmospheric pollution, caused this time by trace amounts of chlorofluorocarbons (CFCs) and some other industrial chemicals, have been reductions of ozone in the high atmosphere and the appearance of so-called 'ozone holes', first over the Antarctic and then over the Arctic polar regions. Ozone effectively filters out ultraviolet (UV) light, and one result of these holes has been increased UV radiation at ground level. UV is high energy radiation with considerable capacity to damage living organisms, particularly by causing lesions in DNA. As outlined in Chapter 2, the spread southwards of the expanding Arctic ozone hole has been implicated in declines of amphibians living at high altitudes in North America and which lay their eggs in shallow water exposed to the sun (Blaustein *et al.*, 1994). There are many amphibian species, particularly palaearctic anurans, which live at high altitude

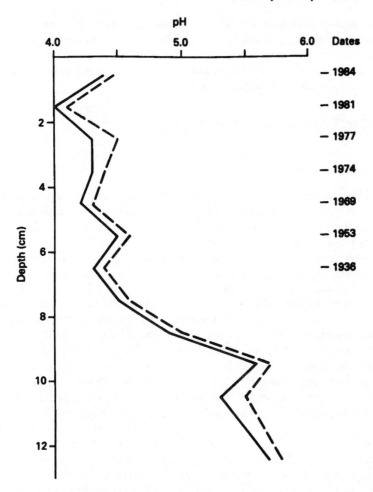

Figure 8.5 Acidification of a natterjack toad (*Bufo calamita*) breeding pond as determined from diatom analysis of a sediment core. Solid and broken lines are two separate computations based on the same data but using different relationship equations. (Reprinted from *Biological Conservation* 53, T.J.C. Beebee *et al.*, 'Decline of the natterjack toad *Bufo calamita* in Britain: palaeoecological, documentary and experimental evidence for breeding site acidification', p. 14 (1990) with kind permission from Elsevier Science Ltd, The Boulevard, Langford Lane, Kidlington OX5 1GB, UK.)

or latitude and habitually expose their eggs in this way to maximize development rates in otherwise cold climates. It remains to be seen how many more, apart from the two identified in the United States, might be suffering from this unexpected danger. Indeed, the effects of elevated UV irradiation on ecosystems are likely to be complex and far-reaching; thus increases in UV initially depress algal growth in freshwater ponds, but this situation is later reversed because invertebrate consumers of algae (especially chironomids) are also killed, resulting in delayed algal blooms (Bothwell, Sherbot and Pollock, 1994). It looks as if UV could turn out to be a more widespread and serious threat to amphibians than acid rain seems to be.

8.5 DIRECT KILLING BY HUMANS

8.5.1 Human predation

Amphibians are used directly by humans in countries all over the world and for a variety of purposes. Dendrobatid frogs in South America donate their skin toxins to anoint hunting arrows; desert-dwellers such as the water-holding frog *Cyclorana platycephalus* are dug up by thirsty aboriginals in Australia; but undoubtedly the most widespread exploitation is as a source of food. A range of species, especially the larger ones, end up in the cooking pot: everything from the giant salamanders (*Andrias* species) of east Asia to the large leptodactylids (such as *Leptodactylus fallax*, served up as 'mountain chicken') of Central and South America are hunted for this purpose, for the most part probably with little effect on population sizes as long as only local economies are involved.

Unfortunately an altogether different situation has arisen with the development of frogs' legs as a delicacy in parts of the western world. Originally confined mostly to green frogs in mainland Europe, rising demand together with increased protection of amphibians in that continent has generated massive export trades, initially in green frogs from eastern Europe but later in 'replacement' species such as the Indian bullfrog *Rana tigrina* from further afield. By the mid-1980s Bangladesh, India and Indonesia were each exporting some 3 million kg of frogs' legs every year, representing massive culls with unknown consequences for the species involved. India banned frog exports in 1987, and Bangladesh followed suit in 1992 after studies indicated that exporting 50 million frogs a year had left only 400 million in the country and the paddy fields depleted of a free, natural insecticide. Indeed, the negative correlation between decreasing frog numbers and increasing purchase of expensive chemical pesticides in Bangladesh provided a stark economic case for ending the trade, but elsewhere it continues with increasing concern that the predation is often well above sustainable yields. The developed world currently consumes a massive 6500 tons of frogs' legs every year (it takes at least 12 legs to make a meal), and until this demand is controlled (or preferably removed) the pressure on palatable species will remain high. There seems little excuse for perpetuating this barbarous practice (the frogs are commonly killed just by cutting them in half) in a so-called civilized society.

Amphibians are also collected for research, education and simply to be kept as pets, and a commercial trade exists to service these needs. Numbers are, at least in most cases, probably too low to warrant serious concern but this is not always the case. The common frog *Rana temporaria*, for example, has a long history of commercial exploitation in Britain with a recent usage of 50 000–100 000 per year. Frog sales have declined steadily in recent decades and by 1988 were below 40 000, though about half of these were taken from a single area (Cornwall) where frogs are abundant. This apparently quite high localized collection rate actually amounted to less than 3% of the adult population, and although probably sustained for several decades, there is no evidence that frogs have declined in Cornwall as a result (Cooke, Morgan and Swan, 1990). On the other hand, the recovery in numbers of a frog population in the midlands of Britain, where amphibians had declined markedly because of habitat destruction, was associated with a decreased rate of spawn collection from 35–50% per annum in the 1970s to less than 25% by the early 1980s (Cooke, 1985). Collection from small or isolated populations may therefore be more serious than large-scale abstraction from big thriving ones, but more evidence is badly needed on this issue.

8.5.2 Accidental killing

One aspect of human life style which impacts conspicuously on amphibians is the driving of motor vehicles. Hundreds of thousands, if not millions, of amphibians are probably killed by motor traffic every year over the world as a whole. This happens most dramatically when busy roads are crossed by animals migrating to breeding sites, but in many countries there is a steady attrition rate throughout the year because amphibians find warm, wet roads attractive places to sit or hunt on after dark. The carnage is appalling, and may have unpleasant consequences for drivers as well as their victims; cases are on record of people being killed when cars have run out of control after skidding on squashed amphibian corpses.

Road mortality of amphibians raises two distinct issues: is it a serious threat to populations, and is it in any case morally acceptable? The former cannot be answered unequivocally because it depends upon both the species involved and the volume of traffic at particular sites. The survival rates of slow-moving salamanders in the United States may be so low as to cause local extinctions when they have to cross roads with only moderate traffic density. In Europe the common toad *Bufo bufo* has received more study than any other species in this context and from 4% to perhaps 50% of adults, depending on traffic volume, may be killed when crossing roads during spring migrations to breeding ponds. Despite this, it is remarkable how long toad populations persist and return to cross the roads (or die on them) year after year and local extinction of this species as a result of traffic mortality alone seems to be rare. However, the killing of wildlife on roads must surely be regarded as morally reprehensible and everything reasonable should be done to minimize it.

How to conserve amphibians

9.1 GENERAL STRATEGIES

Given the threats to amphibians outlined in Chapter 8, what are the best courses of action to ensure their futures? A first step must be to assess levels of risk and prioritize species, as far as knowledge permits, with respect to conservation needs. Globally this has been attempted by the IUCN Red Data Book listings, and by allocation to CITES Appendices. In 1988, the Red Data Book listed 30 anurans, 24 urodeles and no caecilians (i.e. < 1%, < 7% and 0% of all known species) while CITES Appendices contained 56 anurans but only 3 urodeles (< 2% and about 1.3% of all known species) and again no caecilians (Groombridge, 1988). The discrepancies between these two appraisals highlight the primary difficulty in even agreeing upon what the priorities are, but these attempts are constantly revised and updated, and also developed locally as well as globally. At the end of the day, countries or other political units (such as the Council of Europe, or the European Union) decide upon plans of action on the basis of as wide a range of information as possible, very often with local bias of varying degree. A European Red List in 1991, for example, included seven anurans and 12 urodeles but only two each of these featured in the IUCN or CITES global listings.

Given that vulnerable species are identified, the question then becomes one of how best to act. Six critical aspects of a comprehensive conservation strategy can be identified, notably legislation, site acquisition, applied research, habitat management, captive breeding and translocation. All of these can be important, and will be discussed in turn.

9.2 LAWS PROTECTING AMPHIBIANS

Effective legislation is an essential backdrop to all conservation, providing as it does an ultimate sanction against recalcitrance, ignorance and malice towards wildlife and its habitats. Many of the most important laws, either in place or still needed, do not relate to amphibians directly; for instance, decent regulations concerning the use of agrochemicals and the control of atmospheric pollution are critical to amphibian conservation but are usually thought of in the broader context of environmental protection. Nevertheless, herpetologists would do well to add their collective voices towards improved legislation in both these areas.

Laws designed specifically to protect amphibians have been enacted in many countries around the world and details are kept by the IUCN Environmental Law Centre in Bonn, Germany. Examples of such laws operative in 1994 in Australia, countries of the European Economic Union and the United States are given in Tables 9.1 to 9.3, respectively. These are not comprehensive (other US states, apart from those listed, also protect their herpetofauna) but represent those on the IUCN database. Of course they offer only a snapshot of such legislation, which inevitably changes over the years, and do not give a historical perspective because old laws are often subsumed into new ones. In Britain, for example, amphibians were first protected under the Conservation of Wild Creatures and Wild Plants Act of 1975 but this was essentially incorporated, with stronger provisions, into the Wildlife and Countryside Act of 1981. Moreover, the real value of legislation lies in both the detail and the vigour of implementation, neither of which can be gleaned from Act titles alone.

The situation in Australia (Table 9.1) exemplifies what often happens in federal systems, notably that individual states are free to vary the extent to which species are protected. Federal law provides a blanket cover, but is mostly concerned with trade rather than private collection or habitat protection, and some states take matters no further while others have produced much more detailed legislation for amphibian conservation.

Although the European Union is a conglomeration of nation states rather than a federation, one of its aims is to standardize legislation on as many issues as possible. As shown in Table 9.2, it still has a long way to go with respect to wildlife conservation. At one extreme, Portugal had no specific legislation for amphibian conservation at the time of writing, while countries such as Germany, at the other, have long histories of legal protection. Legislation can vary in different parts of the same country, the most dramatic example of this being Italy where there is no federal umbrella and each region goes its own way with spectacularly variable consequences.

Most of the US regulations (Table 9.3) followed the introduction of the much vaunted Endangered Species Act of 1973, and another notable federal law underpins habitat protection for those animals and plants at greatest risk. As in Australia, states vary enormously in subcilliary legislation and while some have none specific for amphibians, others such as California and Texas have protected many or all of their native species.

Laws of this kind are obviously important but often look better on paper than in the field. Total protection rarely means that; apart from the fact that exemptions are almost always possible under licence, enforcement is the ultimate test and this is generally poor. It is inherently very difficult to police a defence of wildlife, and all too often the bureaucracy set up to do so impacts more upon legitimate interests (amateur and professional herpetologists, naturalists and scientists) than on any villains who might be out there. A more important point is that the most serious threats to amphibians, notably habitat destruction and pollution, are addressed only weakly or not at all by the majority of these regulations. Species protection laws essentially relate to direct human predation, a problem that seems usually to be of negligible importance because its scale is so small (with one or two exceptions, such as the market for frogs' legs) compared with the others that amphibians face. Arguably the value of these rules is more to do with education than with the law courts; publicizing how endangered a species is by giving it legal protection can have useful knock-on effects, making landowners, planners and developers more sensitive to the needs of the animals and their habitats. This can facilitate conservation of both protected and unprotected species (e.g. Oldham and Swan, 1991), but there is a delicate balance to be drawn between exploiting endangered status in this way and 'crying wolf' too often, for relatively widespread animals, with the risk that pressure will eventually mount to deregulate all but the most seriously endangered species.

Fortunately most countries also have, or are developing, some level of habitat protection but this has lagged behind species protection because it is politically more difficult to achieve. Prohibiting the occasional capture of a few frogs is one thing; telling landowners what they can or cannot do on their properties is quite another. Europe again provides a good example of the range of habitat protection measures currently available (Corbett, 1989). In most countries, National Parks remain the sole or major form of habitat protection and any value for amphibians is coincidental but can be high, as in the Coto Donãna of southern Spain. France has *Arrêtes de biotope* – areas which can be protected by the Préfet more easily than by acquisition as nature reserves. Ireland has Areas of Scientific Interest (ASIs) which can be designated by government but have little force in law. Some countries do better: in Denmark, for example, the law on nature protection prohibits the destruction of several habitat types, including small ponds and areas of heathland greater than 1 ha in extent. Germany has the potential for legal protection of most biotopes following the Federal Nature Conservation Act of 1987, but action is left to the states (Länders) and only a few places have been safeguarded thus far.

In Britain, although the 'sheltering place' is supposedly protected as well as the species given maximum protection under Schedule 5 of the Wildlife and Countryside Act, the mainstay of habitat conservation in the United Kingdom is the Site of Special Scientific Interest (SSSI), a designation made by government and which is strong enough to result in compulsory purchase if need be. There are several thousand SSSIs, accounting for some 8% of the UK land area, and they include (by accident and design) a considerable number of important amphibian populations. Unfortunately a considerable proportion, probably around 5%, is damaged or destroyed every year despite their legal status (Rowell, 1991), and there are difficulties both in monitoring so many sites adequately and in invoking the legal mechanisms which theoretically protect them.

Table 9.1 Laws for the protection of Australian amphibians

Name of law	Date enacted	Type of protection	Species covered
Wildlife Protection (Regulation of Exports & Imports) Act	1974	Regulation of trade	All native species
National Parks & Wildlife Act	1974, New South Wales only	Total protection, including trade	Species at Governor's discretion, but especially: *Assa darlingtoni, Lechriodus fletcheri, Litoria brevipalmata, L. flavipunctata, L. jervisiensis, L. maculata, L. pearsoniana, L. subglandulosa, Kyarranus loveridgei, K. sphagnicolus, Pseudophryne australis, P. corroboree.*
Territory Parks & Wildlife Conservation Act	1976, Northern Territory only	Total protection, including trade	All native species & introductions pre-1788
Territory Wildlife Regulations	1987, Northern Territory only	Total protection (strengthened) including trade	Protection for all frogs, notably: *Arenophryne rotunda, Cophixalus concinnus, C. saxatilis, Cyclorana vagitus, Litoria longirostris, L. pessonata, L. splendida, Phitoria frosti, Rheobatrachus silus, Uperoleia orientalis.* Prohibited entry: *Bufo marinus.*
Inland Fisheries Regulations	1973, Tasmania only	No liberation allowed	All species
Wildlife Conservation Act	1950, Western Australia only	Partial or total protection	All frog species
Wildlife Conservation Regulation	1970, Western Australia only	Trade regulation	Ranidae and Salamandridae

Table 9.2 Laws for amphibian protection in the European Union

Country	Law	Date	Type of protection	Species covered
Belgium	Ordonage relative à la conservation de la faune sauvage et à la chase	1991 (Bruxelloise)	Total protection	All species
Belgium	Order of the Wallon executive concerning the protection of certain species of native wild vertebrates	1983 (Wallone)	Total protection	All species, though *Rana temporaria* and *R. lessonae/esculenta* may be exempted.
Belgium	Royal Decree on protection measures applicable in the Flemish region for certain species of native wild animals not covered under the implementation of Acts and Orders on hunting, fishing and bird protection	1980 (Vlaamse Gewest)	Total protection	All species except *R. temporaria* (partial protection), and *R. lessonae* may be taken in private rearing ponds.
Denmark	Bekendtgorelse om fredning uf krybdyr, padder, hvirvellose dyr, planter m.m	1991	Total protection including trade	All species on Berne Appendix II; for *Bufo bufo, B. calamita, B. viridis, Rana arvalis, R. dalmatina, R. lessonae/esculenta/ ridibunda, R. temporaria, Triturus vulgaris* and *T. cristatus* eggs and larvae may be collected.
France	Order listing protected species of amphibians and reptiles in France	1979	Total protection	All species except *Rana temporaria* and *R. lessonae/esculenta/ ridibunda.*
Germany	Regulation for the protection of wild animal and plant species	1986	Total protection including trade	All species

Greece	Presidential decree relating to the protection of vegetation and wildlife, as well as the establishment of the procedure	1980	Total protection including trade	*Bufo bufo, B. viridis, Hyla arborea, Mertensiella luschani, Rana dalmatina, R. graeca, Salamandra salamandra, Triturus alpestris, T. vulgaris.*
Ireland	Wildlife Act	1976	Total protection	*Bufo calamita (special protection), also Rana temporaria and Triturus vulgaris*
Italy	Legge Regionale (LR) – Norme per la tutela della natura e modifiche alla legge regionale, dicembre 1979, n.78	1981 (Friuli-Venezia Giulia)	Total protection	All *Rana* species
Italy	LR – Norme sulla detenzione, l'allevamento ed il commercio di animal esotici	1990 (Lazio)	Trade control	All exotic species
Italy	LR – Provvedimenti in materia di tutela ambientale ed ecologia	1977 (Lombardy)	Total or partial protection	All eggs and tadpoles and all toads protected; a closed season for collecting *Rana*
Italy	LR – Norme per la conservazione del patrimonio naturale e dell asserto ambientale	1982 (Piemonte)	Total protection	All species except *Rana*, for which a closed season exists
Italy	Legge Provinciale (LP) – Norme per la tutela di alcune specie della fauna inferiore	1973 (Trento)	Total protection	All eggs and tadpoles, and a closed season for *Rana*
Italy	LR – Norme per la disciplina della raccolta dei funghi e per la tutela di alcun specie della fauna inferiore	1977 (Valle d'Aosta)	Total protection	All eggs and tadpoles, closed season for *Rana*

Italy	LR – Norme per la tutela di alcune specie della fauna inferiore e della flora e disciplina della raccolta dei funghi	1974 (Veneto)	Total protection	All eggs and tadpoles, closed season for *Rana*
Italy	LP – Norme per la protezione della fauna	1973 (Bolzano)	Total protection	*Bombina variegata, Bufo bufo, B. viridis, Hyla arborea, Rana esculenta/lessonae, Salamandra atra, S. salamandra, Triturus alpestris, T. carnifex, T. vulgaris*
Luxembourg	Regulation from the Grand Duke prescribing total and partial protection for certain wild animals	1986	Total protection, including trade	All species
Netherlands	Decree on protected indigenous animal species	1973	Total protection including trade	All native species, though excluding eggs and larvae of ranids
Netherlands	Endangered exotic animal species Act	1975	Trade restrictions	*Andrias* species, *Atelopus varius, A. zeteri, Bufo periglenes, B. superciliaris, Dendrobates altobueyensis, D. azureus, Dyscophus* species, *Nectophryniodes, Proteus anguinus, Salamandra atra*

Spain	Real decreto sobre control de los productos afectados por el Acuerdo de Washington, de 3 de marzo 1973, denominado 'Convencion sobre el comercio internacional de especies amenazadas de fauna y flora silvestres (CITES)'	1985	Trade regulations	All amphibians
Spain	Real decreto por el que se regula el catálogo Nacional de especies amenazadas	1990	Total protection including trade	All species except *Rana perezi* and *Salamandra salamandra*
Spain	Descreto por el que se amplia la lista de especies protegidas y se dictan normas para su protección en el territorio de la Communidad Autonóma de Andalucia	1986 (Andalucia only)	Total protection	All species
United Kingdom	Endangered species (import & export) Act	1976	Trade controls	All species except special exemptions (c.30, including *Bufo marinus* and *Rana catesbiana*)
United Kingdom	Wildlife & Countryside Act	1981	Total or partial protection, and restrictions on release of alien species	Total protection for *Bufo calamita* and *Tritiurus cristatus*, partial protection for other native species. Release into wild of *Alytes obstetricans, Bombina variegata, Hyla arborea, Rana esculenta/lessonae/ ridibunda, Triturus alpestris, T. carnifex* and *Xenopus laevis* specifically forbidden.

| United Kingdom | Wildlife (Northern Ireland) Order | 1985 (Northern Ireland only) | Total protection | *Triturus vulgaris* |
| United Kingdom | Wildlife (protection)(Jersey) Law | 1947 | Total protection | *Bufo bufo, Rana dalmatina, Triturus helveticus* |

Table 9.3 Amphibian conservation laws in the USA

Name of law	Region of application	Date	Degree of protection	Species covered
Injurious wildlife	Federal	1974	Controls trade and release into the wild	All species
Endangered and threatened wildlife and plants	Federal	1975	Total protection, including trade	About 20 species including *Ambystoma macrodactylum, Bufo hemiophrys baxteri, B. houstonensis, Eleutherodactylus jasperi, Eurycea nana, Phaeognathus hubrichti, Plethodon nettingi, P. shenandoah* and *Typhlomolge rathbuni*
Designation of critical habitats for fish and wildlife species	Federal	1977	Habitat protection	*Bufo houstonensis, Eleutherodactylus jasperi, Eurycea nana*
Fish, amphibians and reptiles	California	1984	Total protection, including trade	All indigenous species; limit of 4 specimens of any one species for personal possession
Rules relating to endangered or threatened species	Florida	1981	Total protection, including trade	*Hyla andersonii, Rana areolata*

Endangered or threatened plants and animals (rules)	Iowa	1984	Total protection	*Ambystoma laterale, Necturus maculosus, Notophthalmus viridiscens, Rana areolata*
Official State list of endangered vertebrates	Mississippi	1984	Total protection, including trade	*Aneides aeneus, Eurycea lucifuga, Gyrinophilus porphyriticus*
Listings of endangered species and subspecies of New Mexico	New Mexico	1983	Total protection, including trade	*Aneides hardii, Bufo alvarius, B. boreas, Hylactophryne augusti, Plethodon neomexicanum*
Hunting; fishing licences	Ohio	1984	Partial protection	All frogs; licence required to use as bait
Reglamento para regir el manejo de las especies vulnerables y en peligro de extinción en la Estado libre associado de Puerto Rico	Puerto Rico	1985	Total protection, including trade	*Eleutherodactylus eneidae, E. jasperi E. karlschmidti, Peltophryne lemur*
Non-game and endangered species list	South Carolina	Not dated	Total protection, including trade	*Hyla andersonii Plethodon websteri*
Regulations for taking, possessing and transporting protected non-game species	Texas	1977	Total protection, including trade	17 native species and control of *Bufo marinus*
Endangered and threatened species regulations	Wisconsin	1979	Total protection	*Acis crepitans, Ambystoma tremblayi*

Regulations for importation, possession, sale and disposition of live wildlife and exotic species	Wyoming	1976	Partial regulation	Trade and release of *Bufo marinus* and *Xenopus laevis* prohibited

Most recently, the European Union has attempted to strengthen habitat protection in member states by a Habitats Directive, which will permit designations of Special Areas for Conservation (SACs); it remains to be seen whether the political will exists for this to make a real difference, but 11 urodeles and eight anurans are listed under its Annex II as species whose conservation requires the designation of SACs. A general problem with amphibian habitat conservation, however, is that the protection of large numbers of small sites (such as single ponds and their immediate environs) would be the most effective strategy for many species but few countries have yet taken this point on board.

9.3 SITE ACQUISITION

Irrespective of any other form of action, the acquisition by conservation organizations of sites with endangered species is a high priority in almost any circumstance. Although there have been protracted arguments about optimal sizes and shapes of nature reserves, largely based on island biogeography and metapopulation theories, in practice what usually happens is that whatever is available gets leased or purchased; it is a seller's market. A more significant dilemma is that sites with rare species are often not those with highest overall biodiversity (Prendergast *et al.*, 1993), which means that different interest groups may have different priorities. Thus nature reserves obtained by the state, or even the main non-government organizations involved in nature conservation, sometimes have good amphibian populations but often do not include the best that are known. Increasingly, herpetologists (via their societies) should look to obtaining their own reserves while at the same time lobbying governments to designate more reserves with the best amphibian sites. This is of course already happening, and (for example) the British Herpetological Society leases or owns several important sites in the United Kingdom while the Societas Europaea Herpetologica is pressing for the creation of so-called Biogenetic Reserves in European localities with especially diverse herpetofauna assemblages, particularly in East Sardinia and Evros, Greece (Corbett, 1989).

However much effort is put into formal habitat protection, it is clear that only a small fraction of the land of any country will ever be under the direct control of, or strongly influenced by, conservationists. Although this may eventually include most sites for the rarest species, the future of the majority of amphibians will always depend on the fate of the wider countryside and it will be increasingly important to devise methods for conservation that will work in this broader arena. It is here that liaison between statutory conservation bodies and their counterparts in government departments of agriculture, defence, transport and industry will be increasingly critical.

9.4 APPLIED RESEARCH

An immediate question that very often follows the realization that a species is endangered or in decline is: do we know enough about it to save it? The problem can even become acrimonious, with discord between those wanting to do something (anything!) immediately and others preferring a delay for research, to ensure that what is eventually done is likely to work. Of course some actions can be taken without any research; sites can be designated for protection, and overt, damaging developments resisted. No scientific study is required to show that building on the breeding pond of a rare species is not a good idea. On the other hand, mistakes can be made in devising conservation management for species in the absence of adequate knowledge about their specialized requirements. Early efforts to conserve *Bufo calamita* in Britain were confounded by both over-deepening of ponds and translocations to sites where pools were too acid for spawn to survive – errors which preliminary research might have avoided. There are also instances of decline, such as those of the golden toad *Bufo periglenes* in Costa Rica, which are still quite baffling (Pound and Crump, 1994) and which cannot be addressed at all without scientific investigation.

Fortunately, in many cases conservation management and applied research can progress hand in hand. There are four general areas in which research is almost always at least useful, if not essential. Firstly, survey to extend knowledge of distribution can and often does reveal previously unknown sites and thus widens the scope for future conservation. It is sometimes astonishing how long an endangered species can hide; even after 25 years of study in the relatively crowded British Isles, new sites for *Bufo calamita*, for which there are fewer than 50 localities altogether (Figure 9.1), still occasionally turn up (Banks, Beebee and Cooke, 1994). In the case of more widespread species nevertheless considered to be in decline, such as *Triturus cristatus* in Europe, there will be scope for survey to find the best of what must be tens of thousands of sites for many years ahead. Survey for amphibians is currently very advanced in only a few countries (among which Switzerland in particular stands out), quite well developed in rather more, but verging on absent in the majority. The improvement of national and international databases of amphibian sites will therefore be a priority for the foreseeable future.

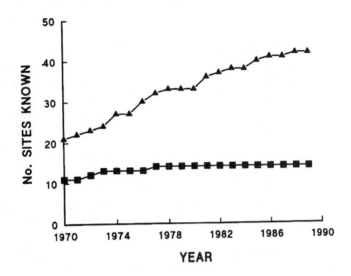

Figure 9.1 Numbers of known rare species (*Bufo calamita*) sites increasing with time: ▲ = total sites in Britain; ■ = total sites in Britain excluding Cumbria county. (Reprinted from *Biological Conservation* 67, B. Banks, T.J.C. Beebee and A.S. Cooke, 'Conservation of the natterjack toad *Bufo calamita* in Britain over the period 1970–1990 in relation to site protection and other factors', p. 113 (1994) with kind permission from Elsevier Science Ltd, The Boulevard, Langford Lane, Kidlington OX5 1GB, UK.)

Secondly, monitoring methods may need development. If the results of conservation management are to be assessed it is crucial to know how populations are responding, and this requires techniques that are usually compromises between accuracy and simplicity, such as counting spawn clumps (Figure 9.2), or just numbers of congregating animals, in breeding ponds. Monitoring methods often need tailoring to the biology of particular species, and are not always transferable even to closely related amphibians; counting spawn clumps, for example, is a reasonable method for estimating population sizes of brown frogs such as *Rana temporaria* but cannot be applied to green frogs like *R. ridibunda* because they lay multiple small clutches hidden in vegetation.

Beyond these two basic needs, it is normally desirable to improve autecological knowledge of the species in question and thus understand, in as great detail as possible, its habitat requirements. These can be much more subtle, for instance with regard to pond permanence or terrestrial habitat structure, than is immediately apparent. Finally, proper scientific tests of the efficacy of any management methods decided upon should be made as a (hopefully) final check that deductions from autecology were correctly made.

Figure 9.2 Mats of frog (*Rana temporaria*) spawn from which numbers of adult females can be estimated. (Photo: A.S. Cooke.)

Other types of study, such as into the degree and type of genetic variation present within and between isolated populations, will also be important in some instances. Scientific research is inevitably an extra expense and like any other aspect of conservation the need for it should be critically assessed in each particular case. There is no doubt that it is often either necessary or desirable, and temptation to dispense with it not infrequently amounts to a false economy.

9.5 HABITAT MANAGEMENT AND CREATION

Amphibians decline because their world changes; sometimes there is simply less of it, but often the problem is about alterations in land-use or the impact of atmospheric pollutants. In these situations site acquisition is not of itself enough; attempts must be made to rejig the past, and the habitat managed in ways which will again allow amphibians to prosper. A common requirement is to reverse the loss of wetland breeding habitat by the restoration of old ponds or the provision of new ones. This important type of management is rarely difficult, and with mechanized excavators can be carried out quickly and on a large scale (Figure 9.3). Needless to say, it generally benefits many organisms apart from amphibians and pond creation and maintenance are major tools in the conservationist armoury.

Figure 9.3 Pond excavation to create new habitat for crested newts *Triturus cristatus*. (Photo: T.J.C. Beebee.)

It is important to realize, however, that there is more to successful pond creation than just digging a hole in the ground. Shape, depth and substrate can all be critical and, for example, there are many species of amphibian that will only breed successfully in temporary ponds that dry out in a summer of average rainfall. For these animals, too much water is every bit as unsatisfactory as too little and creating the right pond becomes something of an art form. Restoration of ponds may include dredging out accumulated silt and removing overgrowth or shading, but in some cases there should also be more subtle work such as the elimination of predatory fish or the reversal of acidification from atmospheric pollutants. Fish can be killed off by temporary drainage or by the use of selective biodegradable piscicides such as rotenone, while acidification problems are soluble in several ways. Best of all, and probably the only long-term hope, is the reduction of the polluting gases. In the meantime, addition of $CaCO_3$ or $Ca(OH)_2$, at around 0.25 kg/m^2 water surface area, usually suffices to restore pH to circumneutral values and has been carried out as a management palliative in many ponds and lakes in the northern hemisphere. Although pH and many other properties return to pre-pollution levels after liming, not all nutrients are restored (Broberg, 1987) and the process has to be repeated year after year. Liming does greatly improve matters for amphibians, however (e.g. Beattie, Aston and Milner, 1993). Another option, for shallow heathland ponds, is the scraping away of bottom sediment with its accumulated acid load; this too raises pH to pre-pollution levels, but is of course immediately vulnerable to further pollution if emissions remain uncontrolled.

Although it is normally preferable to utilize the natural water table in pond creation or restoration, artificial liners such as plastics and concrete are also available and have proved very successful in some situations. Concrete, for example, effectively neutralizes acid precipitation and can provide excellent breeding ponds for species living in vulnerable habitats.

Terrestrial as well as aquatic habitats often need management attention. At one extreme, the removal of woodland, scrub, rank grassland and similar vegetation structures by intensive farming can reduce or eliminate amphibian populations (e.g. Beebee, 1977b), and in such circumstances areas of natural vegetation of as large an extent as possible should be allowed to grow back. In the case of pond-breeding species, this terrestrial habitat should be restored around or very near the pools. At the other extreme, some sub-climax terrestrial habitats such as dunes and heaths have become overgrown due to the cessation of historical farming methods, especially livestock grazing. This is highly damaging to species that require open terrain, and efforts in this situation are directed towards scrub control or removal. Good results can be obtained using various combinations of physical clearance (by hand-cutting or machine: Figure 9.4) and selective herbicides, but this kind of management is expensive and time-consuming and unless followed up in other ways will need regular repetition (e.g. Marrs, 1984, 1987). Probably the ideal plan, wherever possible, is to return to whatever were the historical management techniques and, on heath and dunes in particular, to restore low-density livestock grazing regimes. This kind of strategy works out cheaper and is perhaps even profitable eventually (Kottmann *et al.*, 1985); it is in any case more desirable than methods employing pesticides and which are quite unpredictable in their long-term consequences.

Another form of management likely to assume increasing importance is the protection of amphibians from road traffic. Reducing the toll on amphibians has become an increasingly important issue in many countries, and several methods have been tried. Road signs warning motorists of places where large numbers of amphibians are likely to occur have become commonplace, but it is not clear that these have much effect by themselves. In Britain alone there are now several hundred registered 'toad crossings', places where volunteers turn up on spring nights to carry amphibians across the roads. This approach requires organization indefinitely into the future and takes no account of problems faced by toads moving away from the ponds later in spring, or of toadlets dispersing in summer. A more sophisticated but expensive solution is the installation of 'toad tunnels' – concrete tubes through which animals can move safely below the road surface (Figure 9.5). These structures need to be as wide as possible (at least 20–30 cm diameter), of special material (ACO polymer) and with air gaps opening into the road all the way along the top. Special provisions are necessary to reduce the risk of flooding and oil or salt pollution. Robust, specially designed fencing is needed on both sides of the road to funnel migrating animals coming to and leaving the pond towards the tunnel entrances (Langton, 1989). Despite the costs, tunnel systems of this kind are gaining wide acceptance by planners and are steadily increasing in numbers as a result.

(a)

(b)

Figure 9.4 Scrub clearance on sand dunes for *Bufo calamita* conservation. (a) An area before clearance; (b) a nearby area after clearance. (Photos: T.J.C. Beebee.)

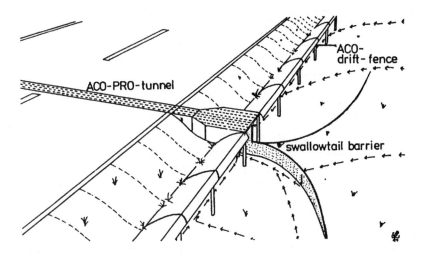

Figure 9.5 Amphibian road tunnel and fence system. (Reproduced from Langton (1989), *Amphibians and Roads*, with permission from ACO Technologies plc.)

There are situations in which tunnels cannot be built and another solution, which may in some cases be better, is to remove the need for road crossing altogether by the provision of a new breeding pond on the other side. This must still be accompanied by fencing on both sides of the highway to prevent animals trying to cross, and effectively either moves the focus of a population or splits it into two. Within four years of such an operation in Germany, migration of *Bufo bufo* towards the original breeding site had dropped to less than 1% of its former rate and the new pond was fully adopted by young and old adults alike (Schlupp and Podloucky, 1994). Of course there are many situations in which this method will also not be possible, and there are concerns about the genetic implications of fragmenting populations into small isolated units (e.g. Madler, 1984; Reh and Seitz, 1990).

Thus, effective methods are available to minimize the massacre of amphibians that occurs on all too many roads every year, and it requires only the political will to implement them more widely. It seems likely that we shall witness substantial proliferations in the use of amphibian tunnels and fencing in the coming years. Wherever practicable, management strategies should aim to maintain metapopulation structures and thus the interconnection of breeding ponds with belts or corridors of suitable terrestrial habitat, including when necessary road tunnel and fencing schemes, so that adult and juvenile amphibians can move and intermix freely over large areas.

Positive steps to create new amphibian habitats, in compensation for those irretrievably lost, should also be entertained as often as possible. The promotion of garden ponds is an outstanding example of how large numbers of small-scale operations can, overall, yield substantial conservation gain. In the wider countryside, ponds can be made in areas previously lacking them and derelict or set-aside farmland converted into valuable habitats. Colonization of such new sites may occur naturally or may require translocation of spawn or adults, depending on their degree of isolation.

9.6 CAPTIVE BREEDING

An instinctive reaction to the plight of a seriously endangered species is to take a few individuals into captivity and establish a captive-bred line, primarily as an insurance policy against total catastrophe in the wild and so that animals can be released again at some future time when conditions are right. The difficulties of this approach are now well understood: the animals may fail to breed success-fully, and even if they do, there are genetic dangers of inbreeding depression or unintentional selection for survival under captive conditions. The first problem, happily, rarely arises with amphibians. The overwhelming majority of species breed readily in captivity, under simple conditions that are cheap and easy to provide. Genetic constraints are not insuperable, especially if stock are inter-changed between different colonies at intervals to prevent random fixation of alleles. Although it is impossible to be sure that captivity will not of itself select for traits that might be disadvantageous in the wild, it is relatively easy to keep most species in vivaria that mimic the physical features of natural habitats to a very high degree. It is less easy, of course, to mimic predation and other com-plicating factors and it is the lack of these selective pressures that could be seri-ous in long-term maintenance of populations over many generations. Occasional infusion of wild stock, where possible, should minimize this risk.

Captive breeding seems a sensible stratagem for amphibians when circum-stances justify the costs involved, and has been used successfully with a num-ber of very rare species (such as *Bufo houstonensis* in North America, and *Alytes muletensis* in Europe). Indeed, the simplicity and success of captive breeding can even be an embarrassment with highly fecund species, where annual production of many thousands of juveniles can be hard to cope with. Adult amphibians and anuran larvae are quite easy to keep and feed, but urodele larvae and juveniles of all kinds, requiring enormous amounts of tiny live prey, are rather a different matter. Planners of amphibian captive breeding programmes would do well to take account of the consequences of too much success before being faced with the problems it can generate.

9.7 TRANSLOCATION

An important part of any conservation strategy is to maintain as far as possible the known historical range of a species, partly because this is likely to maxi-mize genetic diversity but also as a service to people wishing to see wildlife and who might otherwise have to travel long distances to satisfy their interest. If this cannot be achieved by safeguarding extant sites, perhaps because conservation arrived too late, then attempts can be made to restore the range by the reintroduction to previously occupied sites, or introduction to new sites (collectively referred to as translocation), of viable populations. This strategy will of course also act as a general safeguard by increasing the total number of populations of the species in question. Another, less positive reason for entertaining translocation is by way of rescue, if a population is threatened by development but the prospect exists for moving animals to a suitable site elsewhere.

Translocations, intentional and otherwise, have been a feature of human interference with ecosystems for centuries past and there are many examples of success including dramatic ones such as rabbits and marine toads in Australia. However, rare species seem to be more difficult and a comparative study on attempts with endangered birds and mammals between 1973 and 1986 indicated a failure rate of greater than 50% (Griffith *et al.*, 1989). This analysis highlighted several points likely to be of general importance, including the fairly obvious need for habitat at the receptor site to be of excellent quality, and greater success rates within rather than beyond known historic ranges. Translocations of wild-caught animals were almost twice as successful as those using captive-reared individuals, and there were threshold numbers (which differed between species) below which the attempt was unlikely to work, but above which more animals made little difference.

Translocations of amphibians have been under way for at least 20 years, but examples are still few and opinions differ on how successful they have been. Dodd and Seigel (1991) claimed total failure among the instances they reviewed, but this is surely too pessimistic and Burke (1991), mainly considering results from non-endangered species, took a more favourable view. Large populations of common frogs *Rana temporaria* and toads *Bufo bufo* have been established by translocation in Britain following rescue operations (Cooke and Oldham, 1995) so moving animals to new sites can work, probably with elevated chances of success if extensive autecological research is carried out beforehand. Success, by common consent, means the establishment of a self-sustaining population but there is much less agreement about how this should actually be measured. With the rare *Bufo calamita* in Britain, success of translocations has been widely taken to mean the appearance of a second generation of breeding adults, the first generation being those individuals arising from spawn or tadpoles used in the initial translocation. This takes at least five years altogether, and by this criterion at least four out of five translocations initiated between 1980 and 1985 succeeded.

Guidelines for amphibian translocations can be proposed on the basis of experience thus far:

1. The cause(s) of decline of the species should be understood.
2. The habitat requirements of the species should be understood.
3. The reason why the proposed receptor sites are not currently occupied should be understood (e.g. habitat only recently become suitable, or recovered from serious damage).
4. The receptor sites should be suitable in every way, i.e. within the historical range, with appropriate habitat, legally secure (e.g. as nature reserves) and without atypically high numbers of predators or competitors.
5. The donor material should preferably be spawn or tadpoles in the case of pond-breeding species, with as wide a genetic mix as possible but all from within a single donor population (to minimize the risk of outbreeding depression).
6. The donor population should be the geographically nearest one in the same habitat type as the receptor site. Alternatively, excess progeny from captive animals can be used as donors.
7. Translocation should be repeated for at least two consecutive years.
8. Translocations should be monitored over many years (> 5) to determine how well, or otherwise, they have fared.

Translocation tends to be a particularly emotive issue in conservation, requiring special efforts to obtain agreements between all parties likely to be affected by the outcome.

—10——————————————

Amphibian conservation in action

10.1 CHOOSING EXAMPLES

Long before there was concern about global amphibian declines, efforts were begun to conserve amphibians in various countries around the world. Serious attempts to do this date back to the 1970s, though there were earlier, more limited affairs. Some of this conservation work has gone on long enough to provide useful examples of what can be achieved, and of problems (anticipated or otherwise) that can arise. Conservation is a valuable test of empirical study because successful management or translocation of a population is usually a strong indicator that the ecology of a species is well understood.

The examples cited below form a far from comprehensive list of long-term amphibian conservation work, but give a feel for what can be done. Mostly they are taken from Europe, and there are two main reasons for this. Firstly, that continent has a long history of agricultural and industrial development now coupled, in many areas, with some of the highest human population densities in the world. Wildnerness is increasingly hard to find, and amphibians (like other wildlife) are under particularly severe pressure there, though many other corners of the globe are now facing similar problems as development continues apace. Secondly, information about conservation work is less widely disseminated than purely scientific studies and details can be difficult to discover; the examples cited below are among those I happen to know about, but there must be many that have never come my way. It is still commonplace to bury years of hard labour in private or unpublished reports, partly because of a desire to keep details of rare species secret, but often because publication is not seen as a high priority. This is unfortunate for two related reasons: firstly, conservation programmes would undoubtedly benefit from retrospective peer review in just the same way that science in general does, giving confidence in their validity; and secondly, many potential users never hear of, let alone get to see, the results of conservation work and may end up continuously rediscovering the wheel. It is certainly my hope that in future much more emphasis will be placed on publication of the results of long-term conservation programmes.

Attempts to conserve amphibians can be classified for convenience into three categories based upon the level of perceived threat. This is worthwhile because methods suitable for the most endangered species may not be appropriate for widespread but declining ones, and vice versa.

10.2 VERY RARE SPECIES

10.2.1 Majorcan midwife toad *Alytes muletensis*

The European herpetological community was both surprised and delighted when, in 1980, a species previously known only from the fossil record was found alive and kicking on the Balearic Islands. The Majorcan midwife toad (*Alytes*, formerly *Baleaphryne*, *muletensis*) was discovered in and around streams running through deep mountain gorges in the northern part of the island (Mayol and Alcover, 1981). Fossils suggest that midwife toads were once widespread on Majorca and Menorca, but disappeared from lowlands of the former and entirely from the latter island several thousand years ago and coincident with the fossil appearance of the snake *Natrix maura* and the large frog *Rana perezi*. Both of these predators of *A. muletensis* are thought to have been introduced to the Balearics by humans, and to have caused the subsequent decline of the midwife toad. The inaccessible ravines where the toads survive today (Figure 10.1) are among the few places that these predators have not reached.

Adult *Alytes muletensis* inhabit and feed in rock crevices, emerging rarely except to deposit their tadpoles in the stream pools. Survey and monitoring of this species is therefore mostly done by looking for and counting larvae, and requires abseiling equipment to descend into the gorges where the animals live. It rapidly became apparent, however, that only six gorges and a few pools on the high plateau above them were occupied by the toads and that there were probably in total no more than 13 breeding populations. By the late 1980s total adult numbers were estimated to be between 1000 and 3000. Such small populations are obviously highly vulnerable to extinction, and apart from possible further incursions by *Natrix maura* and *Rana perezi*, perceived threats included agricultural developments on the plateau above the gorges, with risks of water abstraction or contamination, as well as increased disturbance by tourist climbers (Corbett, 1989).

Several steps have been taken to safeguard the future of *A. muletensis*. The species was given strict protection under Spanish law immediately following its discovery, and funds were provided by the European Union towards land purchase and the establishment of a nature reserve to conserve at least some of the toad's habitat. There is a strong case, on the basis of other fauna and flora, for the designation of a much larger 'Biogenetic Reserve' of more than 26 000 ha in this part of Majorca (Corbett, 1989) but this seems rather a distant hope at the time of writing. A Recovery Plan for *A. muletensis* has been drawn up and coordinated by the Majorcan Conselleria d'Agricultura i Pesca, the aims of which include establishing stable populations in at least 10 gorges, regular monitoring of all populations, and the maintenance of a captive breeding

Figure 10.1 Gorge habitat of the Majorcan midwife toad *Alytes muletensis*. (Photo: S. Bush.)

stock on Majorca. In fact, the Majorcan midwife toad proved particularly easy to breed in captivity and populations were established during the 1980s at Jersey and Stuttgart zoos, and later in Majorca itself. The progeny from these animals were used in a series of translocations, beginning in 1988, to try to establish toad populations in previously unoccupied ponds and gorges following survey aimed at identifying suitable receptor sites. Of eight such sites originally chosen, one definitely failed (probably because too many predators were present) but at least three now have breeding populations and males were heard calling in the other four by 1994 although metamorphs were first released there only in 1992.

On the basis of these results, there is cause for optimism that not only can existing populations of *A. muletensis* be conserved but also the numbers and range of the species can be expanded substantially beyond those of recent history. There is of course a limit to the number of suitable gorges in northern Majorca, but there is less constraint on pond creation on the high plateau and plans are under way to increase further the number of midwife toad populations in this large area, including the use of concrete liners where the substrate is too permeable to retain water. Overall, conservation of *A. muletensis* has been a success story, which has incidentally yielded valuable information on translocation methods since the discovery of the animal such a short time ago (Bush, 1994).

10.2.2 Houston toad *Bufo houstonensis*

The Houston toad (Figure 10.2) is considered to be one of the five most endangered amphibians of the United States (Bury, Dodd and Fellers, 1980). It is closely related to the American toad, *Bufo americanus*, from which it may have been isolated as recently as the end of the last glaciation about 10 000 years BP. The Houston toad has been found within historical times in just three areas of south-east Texas; it has probably disappeared from the Houston district, but populations survive further inland in the counties of Burleson and especially Bastrop. *Bufo houstonensis* inhabits sandy or loamy soils, often those supporting loblolly pine (*Pinus taeda*), and persists only in small populations no larger than a few hundred animals. It is an explosive breeder, assembling usually in permanent ponds during the latter part of February and in March, at which time it is relatively easy to monitor or study (Jacobson, 1989).

Figure 10.2 The endangered Houston toad *Bufo houstonensis*. (Reproduced from Bury, Dodd and Fellers 1980), *Conservation of the Amphibians of the United States – a review*, with permission from the US Department of the Interior, Fish & Wildlife Service.)

The main threat to *B. houstonensis* is undoubtedly habitat loss, and various forms of development have been responsible for the extirpation of the species from the Houston area. In Bastrop, especially the state park, there are still many breeding ponds within the forest but even here there have been recent local extinctions where trees have been felled around lakes for amenity purposes. A further worry is the prospect of destructive hybridization with two other local bufonids, *B. woodhousei* and *B. valliceps*. It is known that *B. houstonensis* × *B. woodhousei* crosses generate fertile hybrids which could swamp the rare species, while *B. houstonensis* × *B. valliceps* crosses could also lead to problems because hybrids of one reciprocal are sterile and the other inviable (Blair, 1972). However, hybridization seems to occur very rarely in the wild, at least in the major surviving *B. houstonensis* area of Bastrop; temporal isolation between the Houston toad and *B. valliceps*, which breeds much later in the spring, is high, and spatial separation performs a similar function between *B. houstonensis* and *B. woodhousei* (Hillis, Hillis and Martin, 1984). The latter species, which prefers open country, can replace *B. houstonensis* in sites where forest is cleared or sand removed, and this breakdown of allopatry does generate at least some natural hybrids.

Bufo houstonensis is protected by both federal and state laws. The species has been the subject of long-term ecological studies, and attempts are in hand to try to establish populations on private land outside the park area (Johnson, 1992). A captive-breeding colony has been established in the Houston Zoological Gardens, and progeny from these and from wild populations have been used in efforts to establish new populations by translocation. The latter, however, has shown few signs of success following massive efforts during the 1980s, particularly on Attwater Prairie Chicken National Wildlife Refuge in Colorado County, Texas. Despite the release of 62 adults, 6985 newly metamorphosed toadlets and 401 384 eggs, distributed among 10 receptor ponds over several years, no more than a handful of calling males and occasional spawn strings were observed by the late 1980s. Reasons for these difficulties are not clear, but predation of tadpoles at the release sites by snakes (*Nerodia erythrogaster* and *Thamnophis proximus*), and of toadlets by ants (*Solenopsis invicta*), seems to have been high (Freed and Neitman, 1988). It remains to be seen whether refinement of translocation strategies will eventually restore at least some of the historical range of *B. houstonensis*, but in any case long-term conservation and even expansion of its existing habitat is obviously of crucial importance, and looks feasible at least in the Bastrop district.

10.3 WIDESPREAD BUT LOCAL (SPECIALIST) SPECIES

10.3.1 Natterjack toad *Bufo calamita*

A mere three species of the genus *Bufo* are indigenous to Europe, and of these the natterjack toad has the smallest distribution. Even so, the species is found across most of the western and north-central regions of the continent and cannot be described as globally rare or endangered. The problem for *Bufo calamita* is that over much of its range outside Iberia it is confined to a few distinctive, and often vulnerable, habitat types. Natterjacks in these northerly regions usually live in sandy, sparsely vegetated biotopes such as sand dunes, heathlands and old sandpits and are therefore widely but locally distributed depending on the abundance of these relatively infrequent landscape features (Beebee, 1983). Over the past 100 years or so natterjacks have declined considerably in the north of their range, particularly in Britain, Sweden, Belgium and parts of France and Germany (e.g. Beebee, 1977a; Brinkmann and Podloucky, 1987; Schlyter, Hoglund and Stromberg, 1991). *Bufo calamita* was listed as one of Europe's most vulnerable amphibian species in a report to the Council of Europe in 1992. The fate of the natterjack has been studied in detail in Britain, where it is estimated that some 75–80% of populations extant around 1900 had disappeared by the early 1970s. Although a few small sites for *B. calamita* have been discovered since that time, and decline since 1970 has been much slower, there are now thought to be fewer than 50 extant native populations of natterjacks in Britain and a total of less than 20 000 adults (Banks, Beebee and Cooke, 1994).

The prime cause of natterjack toad declines has been habitat destruction, especially massive losses of heathland to afforestation, urbanization and agricultural reclamation. Cessation of livestock grazing on many surviving heaths and coastal sand dunes has also resulted in ranker vegetation structures and subsequent replacement of *B. calamita* by the competitively superior *B. bufo* (Denton and Beebee, 1994). As if this was not enough, at least some of the old heathland breeding sites of *B. calamita* have been polluted by acid rain to the extent that the toads can no longer use them successfully (Beebee *et al.*, 1990).

A substantial combined programme of autecological research and proactive conservation, funded mainly by the British statutory conservation agencies, has since the 1970s attempted to arrest and reverse natterjack declines in the United Kingdom. The species has been protected by law since 1975, and by 1994 more than 80% of surviving populations also enjoyed habitat protection at the level of SSSI status including 12 at least partly safeguarded within the confines of nature reserves. Monitoring and management have been extensive, with well over 100 ponds created or improved for natterjacks between 1970 and 1990. An intensive programme of pond creation and scrub clearance increased natterjack numbers more than threefold within 20 years at the single remaining heathland site in southern England, and management on coastal dunes has shown similar successes (Banks, Beebee and Denton, 1993; Smith and Payne, 1980). Over 60% of British natterjack sites had received some form of conservation management by 1992. Several captive-breeding populations have been established, and a programme of translocations was initiated during the 1970s. A Species Recovery Programme for *Bufo calamita*, starting in 1992, accelerated the translocation plans very substantially and by 1994 attempts had been made to establish populations at 13 sites in England. The strategy was to concentrate translocations on heathlands in the south and east of the country, and thus restore the major zone of distribution that had been lost within historical times.

Translocation efforts inevitably take some time to assess, but indications are that while three definitely failed the remainder have either definitely succeeded or look promising at the time of writing. Natterjack translocations have normally used spawn (equivalent to the output of two females per year for two consecutive years, but made up of fragments from several strings to maximize genetic variation) seeded into specially prepared ponds. Concrete saucer pools on heathland sites have proved particularly successful for this species (Figure 10.3).

Figure 10.3 Artificial concrete pond created at the start of a natterjack toad *Bufo calamita* translocation in Britain.

Natterjack conservation in Britain has not been without its mistakes and failures. *Bufo calamita* breeds in shallow, ephemeral ponds and early measures to deepen such pools following reproductive failure in dry summers were counterproductive. Overdeepening increases tadpole predation by invertebrates, and occasional mass mortality of larvae due to early pond desiccation is much preferable to creating permanent ponds in which development ultimately fails in many or most years because of this devastating invertebrate predation and competition from other anuran tadpoles. The first attempts to translocate natterjacks failed because (as was subsequently realized) the heathland ponds chosen were much too acid to support larval development. In general, however, natterjack conservation in Britain has proved increasingly successful as autecological knowledge has advanced and exemplifies the vital link between practical efforts and applied research.

10.3.2 European tree frog *Hyla arborea*

Like *Bufo calamita*, *Hyla arborea* (Figure 10.4) has a broad distribution over much of central and southern Europe and cannot be considered a globally rare or endangered species. In common with most European amphibians it is now legally protected almost everywhere, and can best be described as a weak specialist in the more northerly parts of its range where it is more demanding with respect to habitat requirements than many of the species with which it is broadly sympatric. The European tree frog is a small, bright green and particularly attractive amphibian that lives for much of the year in vegetation varying in height from low bramble bushes to high forest canopies. Most, however, seem to dwell within a few metres of ground level and at least in northern Europe choose particular vegetation structures where thermoregulatory, hydroregulatory and foraging requirements can readily be met. During May and June the frogs migrate, sometimes as far as a kilometre but usually much less, to their preferred breeding ponds. These tend to be open and unshaded, often shallow (or with shallow regions) but with relatively high water quality and abundant growths of macrophytes. Pools in pastures, sand dunes and old pits or quarries are especially favoured in northern climes. The calls of males are distinctive and strikingly loud, which has the benefit that this species is relatively easy to record.

Figure 10.4 Male European tree frog *Hyla arborea* in full croak. (Photo: R. Krekels.)

The widespread feeling that tree frogs had declined substantially in northern Europe since the Second World War led to a symposium at which the fate of this species was the central theme (Stumpel and Tester, 1992). Tree frogs have disappeared from or become much rarer in many parts of Germany, the Netherlands, Denmark and Sweden and perhaps other countries too in recent years. Thus in Germany tree frogs have declined by at least 50% in Lower Saxony, and in Sweden the best known site for the species in northern Europe was threatened by introduced crayfish. Denmark is thought to have lost 90–97% of its tree frog populations since 1945, with an annual rate of attrition of some 6–20% of remaining sites during the 1980s.

There is widespread agreement that habitat destruction has been the primary cause. Important examples of this include agricultural intensification around breeding ponds, drainage, eutrophication and acidification of ponds, use of pesticides, and introductions of fish. In most cases these causes were deduced from circumstantial evidence, but this is nevertheless persuasive and supported by detailed investigations such as those of Fog (1988) on the Danish island of Bornholm. Of particular interest is the significance of metapopulation structures to the persistence of tree frogs, an issue upon which researchers that have studied this animal seem to be agreed. *Hyla arborea* needs, at least for optimal conditions, a specialized terrestrial habitat structure and a range of ponds to choose from. New ponds, presumably with few predators, are especially good for larval development but temporary pools of any age have similar benefits and within any one area the best pond for tadpole survival varies unpredictably from year to year. Such complex habitat structures are particularly vulnerable to damage, with the implication that even the loss of ponds apparently unimportant to tree frogs at a particular time may be disastrous in the longer term.

When this requirement is taken fully into consideration it is clear that conservation efforts for tree frogs can succeed. In the Netherlands, for example, reintroduction of *H. arborea* by release of only about 400 tadpoles over three years at a specially managed site resulted in the establishment of a viable breeding population. In Denmark, many ponds have been restored for this species in provinces such as Bornholm (where perhaps 50% of the country's tree frogs now survive) and Aarhus, in the latter case accompanied by release of thousands of froglets to try to re-establish viable populations. Some of these efforts have attracted widespread publicity, including for example the production of an information leaflet for general circulation in the Vejle district. Important practical considerations for tree frog conservation include the creation and maintenance of a network of open, shallow ponds without too much shading vegetation but within reach of (and well connected to) suitable terrestrial habitat structures such as stands of bramble bushes. Eutrophication of ponds, and introduction of fish, must be avoided. Although *H. arborea* coexists with some anurans (such as *Bombina* species), there are indications that predation or competition by others (such as green frogs, or *Bufo bufo*) may be deleterious and this is an area that warrants more study if conservation management for tree frogs is to be further refined.

10.4 WIDESPREAD GENERALIST SPECIES

10.4.1 Nature of the problem

Many amphibians are able to survive in a wide range of habitat types, including woodland, agricultural and even urban sites – almost anywhere, in fact, provided that there is some permanent cover and, in the case of aquatic-breeding species, one or a few small ponds. These generalists are inherently the most successful species today and do not have the same priority for conservation as the types mentioned above. Nevertheless, ponds in particular are an endangered habitat form and there is increasing interest in broader conservation strategies that will ensure that currently widespread organisms do not continue to decline and thus become the endangered species of the future. Because of the sheer scale and amount of work involved in having an impact on widespread species, the problems for conservationists are different from but at least as difficult as those posed by rarer animals and plants.

One way to tackle this situation is not to focus on particular species, but on habitats or mixtures of species. Thus the legal protection of ponds in countries such as Denmark, Sweden and parts of Germany is bound to help to conserve the widespread species of amphibians that use them and has much to be commended. An alternative is the concept of species assemblages, in which ponds and their environs are rated on a scoring system according to how many amphibian species use them and also the sizes of the populations (averaged over several years) involved. The best sites can then be protected in some way. This approach has been taken in Britain, but of course requires a substantial survey and monitoring effort over many seasons and even then there is no guarantee that protection (in that country as SSSI) will be forthcoming. The simpler and quicker method of general habitat protection would seem the safer way forward. A third option is to focus on particular widespread species thought to be at risk or declining faster than most, and use them as keystones for the protection of habitats in which they live. An example of such a species is described below.

10.4.2 Great crested newt *Triturus cristatus*

The crested newt must rank as one of Europe's most impressive amphibians. Both sexes are relatively large (15–16 cm long) and strikingly marked on their ventral surfaces with orange, yellow and black, while males in the breeding season have prominent dorsal crests as well as conspicuous silver stripes on each side of their tails. *Triturus cristatus* has a wide distribution over most of northern and central Europe, and in spring can be found in a broad range of small or medium-sized ponds in a similarly broad variety of terrestrial habitats. Crested newts turn up regularly in farm ponds, woodland pools, sand dune slacks and old pit or quarry sites, and more occasionally in parks and gardens. This urodele is primarily a lowland species, rarely occurring in mountainous districts but happy enough at moderate elevations up to at least a few hundred metres above sea level. It requires terrestrial habitat with vegetation dense enough to provide cover and prey during the summer months; scrub or open woodland is particularly favoured but rank grassland or similar structures constitute adequate alternatives. Breeding ponds need to be above some uncertain (but quite small) minimum size, preferably mineral-rich and without predatory fish. Large newt populations often occur where pools dry up in some but not all summers, a regime which presumably minimizes the chance of colonization by fish.

Warning signals of possible serious declines in crested newt populations began to flash in the 1970s. In Britain, *T. cristatus* was recognized as the least abundant of that country's generalist species and appeared, on the basis of questionnaire survey, to have declined in post-war years more severely than its congeners though the rate of decline fell substantially, to a loss of perhaps 0.4% of sites per annum, by the early 1980s (Beebee, 1975; Oldham and Nicholson, 1986). Similar concerns were raised elsewhere in northern and central Europe, particularly in Norway and Finland where the species is on the edge of its range, but also in the Netherlands, Germany and Switzerland, ultimately leading first to its inclusion on Appendix II of the Berne Convention and more recently to a comparable listing on the European Union's Habitats Directive. *Triturus cristatus* is now strictly protected by national laws in most of the countries it inhabits.

Reasons for the decline of crested newts seem to be fairly straightforward. Habitat destruction, especially the loss of breeding ponds but sometimes serious damage to terrestrial habitats, is undoubtedly the most important problem. This in turn is primarily due to the combination of urban development and agricultural intensification that has featured so strongly in recent decades over much of the European continent. Pond acidification may also have played a role in some places, and breeding site loss by seral succession following neglect, and as a result of fish stocking, have all contributed to declines of this species. Critical to understanding the predicament of *T. cristatus*, however, is figuring out why this generalist is inherently rarer, and declining faster, than many other species with which it cohabits and which might also be expected to suffer from the same list of woes. Surveys in Britain revealed that crested newts occurred in only about 2% of ponds within their range, whereas smooth newts *T. vulgaris* (with which *T. cristatus* is often found) inhabited perhaps 27%, and *T. helveticus* 17%, of available water bodies (Swan and Oldham, 1993). The more exacting requirements of crested newts with respect to both breeding and terrestrial habitats are probably what make the difference. These are apparently small but important: *T. cristatus* seems to need slightly larger ponds than its congeners with areas of open water in which males can establish temporary territories, and its larvae are more vulnerable to fish predation, a combination which rules out successful use of many small breeding sites (including most garden ponds) in which the smaller newts get along quite well. Crested newts may also need more extensive areas of suitable terrestrial habitat than the smaller species (e.g. Beebee, 1977b) but this is probably less critical than the ponds.

Conservation measures aimed at reducing crested newt declines have been slow to get off the ground – probably because, being so widespread, there is little impetus for priority treatment. As yet, not many countries have even taken the essential first step of surveying and documenting their crested newt populations. This is a daunting task for what is still a very widespread amphibian, but a serious conservation strategy (as opposed to *ad hoc* efforts when opportunities arise) certainly requires a detailed knowledge of distribution and identification of the largest populations. Leaders in this area are Switzerland with over 8000 ponds surveyed (Grossenbacher, 1988), some of the German Landers, and the United Kingdom which during the 1980s organized surveys that ultimately covered some 11 000 water bodies (perhaps 3% of the British total) and in doing so identified more than 3000 ponds used for breeding by crested newts (Swan and Oldham, 1993). From the British survey it is possible to deduce by extrapolation that there are probably of the order of 18 000 breeding sites for *T. cristatus* in the country as a whole, a number which highlights the difficulty of conserving what is still a relatively common animal.

Practical conservation has progressed in the Netherlands, in the absence of a comprehensive survey, by the creation in provincial pond programmes of several hundred new pools mainly in the south and east of the country. In Sweden, again in the absence of systematic survey, there has been an attempt to conserve the species by making it illegal to destroy known crested newt sites or introduce fish into them. The difficulties of enforcing such regulations are not difficult to imagine. In Britain crested newts and their breeding sites are also theoretically safe from destruction under Part I of the Wildlife and Countryside Act of 1981; it is intended to protect the best crested newt sites within SSSIs under Part II of this same Act, but there is as yet no agreement about how many sites should be thus protected (though some already are). The key difficulties are finding out when crested newt sites come under threat (because so many exist), and identifying the best populations when only about one in six of those thought to occur have yet been documented. Population size is accepted as the main criterion for conservation priority and short-lists for site protection amounting to some 50–100 populations (i.e. about 0.4% of the estimated 18 000 British sites) have been identified but as yet received little action. In turn, the 'best of the best' sites throughout the species' range should receive the higher level of habitat protection conferred by SAC status following implementation of the European Union's Habitats Directive.

Until a comprehensive strategy to deal with these problems is developed, crested newt conservation will no doubt continue on an essentially pragmatic basis. There are already some carefully documented examples upon which the implementation of a broader strategy can eventually be based. Extensive consultations with landowners resulted in the saving or better management, by peripheral scrub removal, of more than 10 crested newt ponds in one small part of England (Cooke, 1990). Some open-cast coal mining areas in Britain have been subjected to particularly intensive conservation efforts for *Triturus cristatus*, with the creation of conservation areas, new ponds and even large-scale captive breeding of the newts by the mining organizations, followed by release of larvae into the ponds. Among other things, this programme has shown that seeding a pond with 100–200 large newt larvae can be enough to ensure the return of mature adults 2–3 years later. In general, however, the main consequence of strict protection for crested newts has been improved publicity about the plight of ponds in general and a tendency to rescue animals from threatened sites and translocate them elsewhere rather than to save the site itself. Moving this species around is therefore a relatively frequent event and at the very least it would be nice to know how successful these translocations are. Clearly some of them have worked, but as yet there is too little information to be confidently prescriptive about the types of pond that constitute good receptors. More research on the consequences of these 'experiments' is surely urgent.

Effective strategies for conserving amphibians like *T. cristatus* are thus still in their infancy. One way ahead, which is straightforward in principle, is to aim as a bare minimum to replace the estimated numbers of populations lost each year. This can be calculated quite simply from the total known sites (extrapolated if necessary from survey data) and estimated loss rates; in Britain, for example, losing 0.4% of 18 000 populations infers a need to generate about 70 new ones each year just to stand still. This does not require any knowledge of precisely which populations are going under, and spread across the various regions does not seem an excessively onerous goal.

10.5 OVERVIEW

Practical measures to conserve amphibians are usually possible and often fairly straightforward; mostly they amount to restoring habitats and maintaining them in the conditions of a few decades or more ago. It is not always that easy: pollution of ponds by acid rain requires more complex solutions, and the impact of increased ultraviolet irradiation may be completely intractable in the short-term. There are still species declining or disappearing for reasons currently quite unknown, and for which at present all that can be done is to try to establish captive-breeding populations. In general, though, the armoury of methods available to conservationists works well with this group of relatively sedentary animals. Habitats can be protected and managed, populations can often be re-established by translocation, and most species reproduce without too much difficulty in captivity. As usual in conservation, it is the political will to carry out the necessary measures, with all its implications for public expenditure and aggravation of landowners, that will decide what actually happens. The onus is on all of us to make our views known and to press for action now; if we fail, our descendants will surely judge us harshly.

> The exterminators. I turned the harsh word over in my mind. Great God, what were we doing? The net was drawing tighter. Man had his hands upon it. The effects were terrifying. I thought of a sparkling stream where I had played as a boy. There had been sunfish in it, turtles. Now it ran sludge and oil. It was true. The net was tightening. All over the earth it was tightening.
>
> Loren Eisely, in *The Invisible Island*.

References

Adler, K. (1982) Sensory aspects of amphibian navigation and compass orientation. *Vertebrata Hungarica* **21**, 7–18.

Alcock, M.R. and Mordon, A.J. (1981) The sulphur content and pH of rainfall and throughfalls under pine and birch. *Journal of Applied Ecology* **18**, 835–839.

Alford, R.A. and Harris, R.N. (1988) Effects of larval growth history on anuran metamorphosis. *American Naturalist* **131**, 91–106.

Alford, R.A. and Wilbur, H.M. (1985) Priority effects in experimental pond communties: competition between *Bufo* and *Rana*. *Ecology* **66**, 1097–1105.

Altig, R. and Channing, A. (1993) Hypothesis: Functional significance of colour and pattern in anuran tadpoles. *Herpetological Journal* **3**, 73–75.

Andren, C. and Nilson, G. (1985) Habitat and other environmental characteristics of the natterjack toad, *Bufo calamita*, in Sweden. *British Journal of Herpetology* **6**, 419–423.

Andren, C., Henrikson, L., Olsson, M. and Nilson, G. (1988) Effects of pH and aluminium on embryonic and early larval stages of Swedish brown frogs, *Rana arvalis*, *R. temporaria* and *R. dalmatina*. *Holarctic Ecology* **11**, 127–135.

Andren, C., Marden, M. and Nilsson, G. (1989) Tolerance to low pH in a population of moor frogs, *Rana arvalis*, from an acid and neutral environment: a possible case of rapid evolutionary response to acidification. *Oikos* **56**, 215–223.

Arak, A. (1983) Sexual selection by male–male competition in natterjack toad choruses. *Nature* **306**, 261–262.

Arak, A. (1988) Callers and satellites in the natterjack toad: evolutionarily stable decision rules. *Animal Behaviour* **36**, 416–432.

Arnold, E.N., Burton, J.A. and Ovenden, D.W. (1978) *A Field Guide to the Reptiles and Amphibians of Britain and Europe*. Collins, London.

Arnold, H.R. (1973) *Provisional Atlas of the Amphibians and Reptiles of the British Isles*. Biological Records Centre, Huntingdon.

Arntzen, J.W. and Teunis, S.F.M. (1993) A six year study on the population dynamics of the crested newt (*Triturus cristatus*) following the colonisation of a newly created pond. *Herpetological Journal* **3**, 99–110.

Aubry, K.B., Senger, C.M. and Crawford, R.L. (1987) Discovery of Larch Mountain salamanders *Plethodon larselli* in the central Cascades range of Washington. *Biological Conservation* **42**, 147–152.

Avery, R.A. (1968) Food and feeding relations of three species of *Triturus* (Amphibia, Urodela) during the aquatic phase. *Oikos* **19**, 408–412.

Baker, J.M.R. (1990) Body size and spermatophore production in the smooth newt (*Triturus vulgaris*). *Amphibia–Reptilia* **11**, 173–184.

Baker, J. and Waights, V. (1993) The effect of sodium nitrate on the growth and survival of toad tadpoles (*Bufo bufo*) in the laboratory. *Herpetological Journal* **3**, 147–148.

Baker, J. and Waights, V. (1994) The effects of nitrate on tadpoles of the tree frog (*Litoria caerulea*). *Herpetological Journal* **4**, 106–108.

Banks, B. and Beebee, T.J.C. (1986) A comparison of the fecundities of two species of toad (*Bufo bufo* and *Bufo calamita*) from different habitat types in Britain. *Journal of Zoology (London)* 208, 325–338.

Banks, B. and Beebee, T.J.C. (1987a) Spawn predation and larval growth inhibition as mechanisms for niche separation in anurans. *Oecologia* 72, 569–573.

Banks, B. and Beebee, T.J.C. (1987b) Factors influencing breeding site choice by the pioneering amphibian *Bufo calamita*. *Holarctic Ecology* 10, 14–21.

Banks, B. and Beebee, T.J.C. (1988) Reproductive success of natterjack toads *Bufo calamita* in two contrasting habitats. *Journal of Animal Ecology* 57, 475–492.

Banks, B., Beebee, T.J.C. and Cooke, A.S. (1994) Conservation of the natterjack toad *Bufo calamita* in Britain over the period 1970–1990 in relation to site protection and other factors. *Biological Conservation* 67, 111–118.

Banks, B., Beebee, T.J.C. and Denton, J.S. (1993) Long-term management of a natterjack toad (*Bufo calamita*) population in southern Britain. *Amphibia–Reptilia* 14, 155–168.

Barandun, J. (1990) Reproduction of yellow-bellied toads *Bombina variegata* in a manmade habitat. *Amphibia–Reptilia* 11, 277–284.

Bauer, A.M. (1988) A geological basis for some herpetofauna disjunctions in the southwest Pacific, with special reference to Vanuatu. *Herpetological Journal* 1, 259–263.

Baxter, G.T., Stromberg, M.R. and Dodd, C.K. (1982) The status of the Wyoming Toad (*Bufo hemiophrys baxteri*). *Environmental Conservation* 9, 348.

Beanland, T.J. and Howe, C.J. (1992) The inference of evolutionary trees from molecular data. *Comparative Biochemistry and Physiology* 102B, 643–659.

Beattie, R.C. (1985) The date of spawning in populations of the common frog (*Rana temporaria*) from different altitudes in northern England. *Journal of Zoology (London)* 205, 137–154.

Beattie, R.C., Aston, R.J. and Milner, A.G.P. (1991) A field study of fertilisation and embryonic development in the common frog (*Rana temporaria*) with particular reference to acidity and temperature. *Journal of Applied Ecology* 28, 346–357.

Beattie, R.C., Aston, R.J. and Milner, A.G.P. (1993) Embryonic and larval survival of the common frog (*Rana temporaria*) in acidic and limed ponds. *Herpetological Journal* 3, 43–48.

Beebee, T.J.C. (1975) Changes in status of the great crested newt *Triturus cristatus* in the British Isles. *British Journal of Herpetology* 5, 481–490.

Beebee, T.J.C. (1977a) Environmental change as a cause of natterjack toad *Bufo calamita* declines in Britain. *Biological Conservation* 11, 87–102.

Beebee, T.J.C. (1977b) Habitats of the British Amphibians (1): Chalk uplands. *Biological Conservation* 12, 279–294.

Beebee, T.J.C. (1979) Habitats of the British Amphibians (2): Suburban parks and gardens. *Biological Conservation* 15, 241–258.

Beebee, T.J.C. (1983) *The Natterjack Toad*. Oxford University Press.

Beebee, T.J.C. (1984) Possible origins of Irish natterjack toads (*Bufo calamita*). *British Journal of Herpetology* 6, 398–402.

Beebee, T.J.C. (1985) Discriminant analysis of amphibian habitat determinants in south-east England. *Amphibia–Reptilia* 6, 35–43.

Beebee, T.J.C. (1990) Identification of closely-related anuran early life-stages by electrophoretic fingerprinting. *Herpetological Journal* 1, 454–457.

Beebee, T.J.C. (1991) Purification of an agent causing growth inhibition in anuran larvae and its identification as a unicellular unpigmented alga. *Canadian Journal of Zoology* 69, 2146–2153.

Beebee, T.J.C. (1992) Amphibian decline? *Nature* 355, 120.

Beebee, T.J.C. (1995) Amphibian breeding and climate. *Nature* 374, 219–220.

Beebee, T.J.C. and Wong, A.L.-C. (1992) *Prototheca*-mediated interference competition between anuran larvae operates by resource diversion. *Physiological Zoology* 65, 815–831.

Beebee, T.J.C., Flower, R.J., Stevenson, A.C. *et al.* (1990) Decline of the natterjack toad *Bufo calamita* in Britain; palaeoecological, documentary and experimental evidence for breeding site acidification. *Biological Conservation* 53, 1–20.

Begon, M. (1979) *Investigating Animal Abundance: Capture–Recapture for Biologists.* Edward Arnold, London.

Beiswinger, R.E. (1986) An endangered species, the Wyoming Toad *Bufo hemiophrys baxteri* – the importance of an early warning system. *Biological Conservation* 37, 59–71.

Bell, G. (1977) The life of the smooth newt (*Triturus vulgaris*) after metamorphosis. *Ecological Monographs* 47, 279–299.

Bell, G. and Lawton, J.H. (1975) The ecology of the eggs and larvae of the smooth newt (*Triturus vulgaris* (Linn.)). *Journal of Animal Ecology* 44, 393–423.

Berger, L. (1973) Systematics and hybridisation in European green frogs of *Rana esculenta* complex. *Journal of Herpetology* 7, 1–10.

Berger, L. and Berger, W.A. (1992) Progeny of water frog populations in central Poland. *Amphibia–Reptilia* 13, 135–146.

Berger, L., Uzzell, T. and Hotz, H. (1988) Sex determination and sex ratios in western Palearctic water frogs: XX and XY female hybrids in the Pannonian Basin? *Proceedings of the Academy of Natural Sciences of Philadelphia* 140, 220–239.

Berven, K.A. (1987) The heritable basis of variation in larval development patterns within populations of the wood frog (*Rana sylvatica*). *Evolution* 41, 1088–1097.

Berven, K.A. (1990) Factors affecting population fluctuations in larval and adult stages of the wood frog (*Rana sylvatica*). *Ecology* 71, 1599–1608.

Biesterfeldt, J.M., Petranka, J.W. and Sherbondy, S. (1993) Prevalence of chemical interference competition in natural populations of wood frogs, *Rana sylvatica*. *Copeia* 1993, 688–695.

Blair, W.F. (ed.) (1972) *Evolution in the Genus* Bufo. University of Texas Press, Austin.

Blaustein, A.R. and O'Hara, R.K. (1986) An investigation of kin recognition in red-legged frog (*Rana aurora*) tadpoles. *Journal of Zoology (London)* 209, 347–353.

Blaustein, A.R. and Waldman, B. (1992) Kin recognition in anuran amphibians. *Animal Behaviour* 44, 207–221.

Blaustein, A.R., O'Hara, R.K. and Olson, D.H. (1984) Kin preference behaviour is present after metamorphosis in *Rana cascadae* frogs. *Animal Behaviour* 32, 445–450.

Blaustein, A.R., Wake, D.B. and Sousa, W.P. (1994) Amphibian declines: judging stability, persistence, and susceptibility of populations to local and global extinctions. *Conservation Biology* 8, 60–71.

Blaustein, A.R., Hoffman, P.D., Hokit, D.G. *et al.* (1994) UV repair and resistance to solar UV-B in amphibian eggs: A link to population declines? *Proceedings of the National Academy of Sciences USA* 91, 1791–1795.

Boomsma, J.J. and Arntzen, J.W. (1985) Abundance, growth and feeding of natterjack toads (*Bufo calamita*) in a 4-year old artificial habitat. *Journal of Applied Ecology* 22, 395–405.

Bosch, J. and Lopez-Bueis, L. (1994) Comparative study of the dorsal pattern in *Salamandra salamandra bejarae* (Wolterstorff, 1934) and *S. s. almanzoris* (Muller and Hellmich, 1955). *Herpetological Journal* 4, 46–48.

Bothwell, M.L., Sherbot, D.M.J. and Pollock, C.M. (1994) Ecosystem response to solar ultraviolet-B radiation: influence of trophic level interactions. *Science* 265, 97–100.

Bourne, G.R. (1993) Proximate costs and benefits of mate acquisition at leks of the frog *Ololygon rubea. Animal Behaviour* 45, 1051–1059.

Bradford, D.F., Tabatabai, F. and Graber, D.M. (1993) Isolation of remaining populations of the native frog, *Rana muscosa*, by introduced fishes in Sequoia and Kings Canyon National Parks, California. *Conservation Biology* 7, 882–888.

Bradford, D.F., Gordon, M.S., Johnson, D.F. *et al.* (1994) Acid deposition as an unlikely cause for amphibian population declines in the Sierra Nevada, California. *Biological Conservation* 69, 155–161.

Brinkmann, R. and Podloucky, R. (1987) Vorkommen, Gefahrdung und Schutz der Kreuzkrote (*Bufo calamita* Laur.) in Niedersachsen unter besonderer Berucksichtigung von Abrabungen – Grundlagen fur ein Artenhilfsprogramm. *Ber. naturhist. Ges. Hannover* 129, 181–207.

Broberg, O. (1987) Nutrient responses to the liming of lake Garesjon. *Hydrobiologia* 150, 11–24.

Brodie, E.D. and Formanowicz, D.R. (1981) Palatability and antipredator behaviour of the treefrog *Hyla versicolor* to the shrew *Blarina brevicauda. Journal of Herpetology* 15, 235–236.

Bronmark, C., Rundle, S.D. and Erlandsson, A. (1991) Interactions between freshwater snails and tadpoles: competition and facilitation. *Oecologia* 87, 8–18.

Bruce, R.C. (1976) Population structure, life history and evolution of paedogenesis in the salamander *Eurycea neotenes. Copeia* 1976, 242–249.

Bucci-Innocenti, S., Ragghianti, M. and Mancino, G. (1983) Investigations of karyology and hybrids in *Triturus boscai* and *T. vittatus*, with a reinterpretation of the species groups within *Triturus* (Caudata: Salamandridae). *Copeia* 1983, 662–672.

Buckley, J. (1989) The distribution and status of newts in Norfolk. *Transactions of the Norfolk and Norwich Naturalists Society* 28, 221–232.

Buckley, J. (1991) The distribution and status of the common frog and common toad in Norfolk. *Transactions of the Norfolk and Norwich Naturalists Society* 29, 15–26.

Bull, J.J. and Shine, R. (1979) Iteroparous animals that skip opportunities for reproduction. *American Naturalist* 114, 296–303.

Burke, R.L. (1991) Relocation, repatriation and translocations of amphibians and reptiles: taking a broader view. *Herpetologica* 47, 350–357.

Burton, T.M. and Likens, G.E. (1975) Salamander populations and biomass in the Hubbard Brook Experimental Forest, New Hampshire. *Copeia* 1975, 541–546.

Bury, R.B., Dodd, C.K. and Fellers, G.M. (1980) *Conservation of the Amphibia of the United States: a review*. US Department of the Interior Fish and Wildlife Service Resource Publication No. 134. Washington DC.

Bush, S. (1994) Good news for the Majorcan midwife toad. *Froglog* 10, 3.

Buskirk, J.V. and Smith, D.C. (1991) Density-dependent population regulation in a salamander. *Ecology* 72, 1747–1756.

Carter, R.N. and Prince, D. (1981) Epidemic models used to explain biogeographical distribution limits. *Nature* 293, 644–645.

Cartwright, S.A. (1988) Intraguild predation and competition: an analysis of net growth shifts in larval amphibian prey. *Canadian Journal of Zoology* 66, 1813–1821.

Caughley, G. (1994) Directions in conservation biology. *Journal of Animal Ecology* 63, 215–244.

Champion, A.B., Prager, E.M., Wachter, D. and Wilson, A.C. (1974) Microcomplement fixation, in *Biochemical and Immunological Taxonomy of Animals* (ed. C.A. Wright), Academic Press, London.

Cherry, M.I. (1993) Sexual selection in the raucous toad, *Bufo rangeri*. *Animal Behaviour* 45, 359–373.

Clark, K.L. and Hall, R.J. (1985) Effects of elevated hydrogen ion and aluminium concentrations on the survival of amphibian embryos and larvae. *Canadian Journal of Zoology* 63, 116–123.

Clarke, R.D. (1972) The effect of toe clipping on survival in Fowler's toad (*Bufo woodhousei fowleri*). *Copeia* 1972, 182–185.

Cogalniceanu, D. (1992) A comparative ethological study of female chemical attractants in newts (Genus *Triturus*). *Amphibia–Reptilia* 13, 69–74.

Cook, R.P. (1983) Effects of acid precipitation on embryonic mortality of *Ambystoma* salamanders in the Connecticut valley of Massachusetts. *Biological Conservation* 27, 77–88.

Cooke, A.S. (1970) The effect of pp'-DDT on tadpoles of the common frog (*Rana temporaria*). *Environmental Pollution* 1, 57–71.

Cooke, A.S. (1971) Selective predation by newts on frog tadpoles treated with DDT. *Nature* 229, 275–276.

Cooke, A.S. (1973a) Response of *Rana temporaria* tadpoles to chronic doses of pp'-DDT. *Copeia* 1973, 647–652.

Cooke, A.S. (1973b) The effects of DDT, when used as a mosquito larvicide, on tadpoles of the frog *Rana temporaria*. *Environmental Pollution* 5, 259–273.

Cooke, A.S. (1974) Differential predation by newts on anuran tadpoles. *British Journal of Herpetology* 5, 386–390.

Cooke, A.S. (1975a) Spawn site selection and colony size of the frog (*Rana temporaria*) and the toad (*Bufo bufo*). *Journal of Zoology (London)* 175, 29–38.

Cooke, A.S. (1975b) Spawn clumps of the common frog *Rana temporaria*: numbers of ova and hatchability. *British Journal of Herpetology* 5, 505–509.

Cooke, A.S. (1977) Effects of field applications of the herbicides diquat and dichlobenil on amphibians. *Environmental Pollution* 12, 43–50.

Cooke, A.S. (1981) Tadpoles as indicators of harmful levels of pollution in the field. *Environmental Pollution* 25, 123–133.

Cooke, A.S. (1983) The warty newt (*Triturus cristatus*) in Huntingdonshire. *Report of the Huntingdon Fauna and Flora Society* 36, 41–48.

Cooke, A.S. (1985) The deposition and fate of spawn clumps of the common frog *Rana temporaria* at a site in Cambridgeshire, 1971–1983. *Biological Conservation* 32, 165–187.

Cooke, A.S. (1990) The impact of the Wildlife and Countryside Act, 1981, on the conservation of the crested newt *Triturus cristatus*. *Huntingdonshire Fauna and Flora Society 42nd Annual Report* (1989).

Cooke, A.S. and Frazer, J.F.D. (1976) Characteristics of newt breeding sites. *Journal of Zoology (London)* 178, 223–236.

Cooke, A.S. and Oldham, R.S. (1995) Establishment of populations of the common frog, *Rana temporaria*, and common toad, *Bufo bufo*, in a newly created reserve following translocation. *Herpetological Journal* 5, 173–180.

Cooke, A.S., Morgan, D.H.W. and Swan, M.J.S. (1990) Frog collection with special reference to Cornwall. *British Herpetological Society Bulletin* 33, 9–11.

Corbett, K.F. (1989) *Conservation of European Reptiles and Amphibians*. Christopher Helm, London.

Costanzo, J.P. and Lee, R.E. (1993) Cryoprotectant production capacity of the freeze-tolerant wood frog, *Rana sylvatica*. *Canadian Journal of Zoology* 71, 71–75.

Cronin, J.T. and Travis, J. (1986) Size-limited predation on larval *Rana areolata* (Anura: Ranidae) by two species of backswimmer (Insecta: Hemiptera: Notonectidae). *Herpetologica* 42, 171–174.

Crump, M.L. (1981) Variation in propagule size as a function of environmental uncertainty for tree frogs. *American Naturalist* 117, 724–737.

Cummins, C.P. (1986a) Temporal and spatial variation in egg size and fecundity in *Rana temporaria*. *Journal of Animal Ecology* 55, 303–316.

Cummins, C.P. (1986b) Effects of aluminium and low pH on growth and development in *Rana temporaria* tadpoles. *Oecologia* 69, 248–252.

Cummins, C.P. (1988) Effect of calcium on survival times of *Rana temporaria* embryos at low pH. *Functional Ecology* 2, 297–302.

Cummins, C.P. (1989) Interactions between the effects of pH and density on growth and development in *Rana temporaria* L. tadpoles. *Functional Ecology* 3, 45–52.

Cunningham, A.A., Langton, T.E.S., Bennett, P.M. *et al.* (1993) Unusual mortality associated with poxvirus-like particles in frogs (*Rana temporaria*). *Veterinary Record* 133, 141–142.

Dale, J.M., Freedman, B. and Kerekes, J. (1985) Acidity and associated water chemistry of amphibian habitats in Nova Scotia. *Canadian Journal of Zoology* 63, 97–105.

Davies, N.B. and Halliday, T.R. (1977) Optimal mate selection in the toad *Bufo bufo*. *Nature* 269, 56–58.

Davies, N.B. and Halliday, T.R. (1978) Deep croaks and fighting assessment in toads *Bufo bufo*. *Nature* 274, 683–685.

Davies, N.B. and Halliday, T.R. (1979) Competitive mate searching in male common toads *Bufo bufo*. *Animal Behaviour* 27, 1253–1267.

DeBenedictis, P.A. (1974) Interspecific competition between tadpoles of *Rana pipiens* and *Rana sylvatica*: an experimental field study. *Ecological Monographs* 44, 129–151.

Degani, G. and Hahamou, H. (1987) Enzyme (aldolase) activity in hyperosmotic media (NaCl and urea) in the terrestrial toad, *Bufo viridis* and frog *Rana ridibunda*. *Herpetological Journal* 1, 177–180.

Denton, J.S. (1991) The distribution and breeding site characteristics of newts in Cumbria, England. *Herpetological Journal* 1, 549–554.

Denton, J.S. and Beebee, T.J.C. (1992) An evaluation of methods for studying natterjack toads (*Bufo calamita*) outside the breeding season. *Amphibia–Reptilia* 13, 365–374.

Denton, J.S. and Beebee, T.J.C. (1993a) Reproductive strategies in a female-biased population of natterjack toads, *Bufo calamita*. *Animal Behaviour* 46, 1169–1175.

Denton, J.S. and Beebee, T.J.C. (1993b) Density-related features of natterjack toad (*Bufo calamita*) populations in Britain. *Journal of Zoology (London)* 229, 105–119.

Denton, J.S. and Beebee, T.J.C. (1994) The basis of niche separation during terrestrial life between two species of toad (*Bufo bufo* and *Bufo calamita*): competition or specialisation? *Oecologia* 97, 390–398.

Diaz-Paniagua, C. (1985) Larval diets related to morphological characters of five anuran species in the biological reserve of Donana (Huelva, Spain). *Amphibia–Reptilia* 6, 307–322.

Diaz-Paniagua, C. (1987) Tadpole distribution in relation to vegetal heterogeneity in temporary pools. *Herpetological Journal* 1, 167–169.

Diaz-Paniagua, C. (1988) Temporal segregation in larval amphibian communities in temporary ponds at a locality in SW Spain. *Amphibia–Reptilia* 9, 15–26.

Diaz-Paniagua, C. (1989) Larval diets of two anuran species, *Pelodytes punctatus* and *Bufo bufo*, in south-west Spain. *Amphibia–Reptilia* 10, 71–75.

Diaz-Paniagua, C. (1990) Temporary ponds as breeding sites of amphibians at a locality in southwestern Spain. *Herpetological Journal* 1, 447–453.

Dodd, C.K. (1991) The status of the Red Hills Salamander *Phaeognathus hubrichti*, Alabama, USA, 1976–1988. *Biological Conservation* 55, 57–75.

Dodd, C.K. and Seigel, R.A. (1991) Relocation, repatriation, and translocation of amphibians and reptiles: are they conservation strategies that work? *Herpetologica* 47, 336–350.

Dolmen, D. (1980) Distribution and habitat of the smooth newt, *Triturus vulgaris* (L.), and the warty newt, *T. cristatus* (Laurenti), in Norway. *Proceedings of the European Herpetological Symposium* 1980, 127–139.

Dolmen, D. (1988) Coexistence and niche segregation in the newts *Triturus vulgaris* (L.) and *T. cristatus* (Laurenti). *Amphibia–Reptilia* 9, 365–374.

Dolmen, D. and Koksvik, J.I. (1983) Food and feeding habits of *Triturus vulgaris* (L.) and *T. cristatus* (LAURENTI) (Amphibia) in two bog tarns in central Norway. *Amphibia–Reptilia* 4, 17–24.

Dournon, C. and Houillon, Ch. (1985) Thermosensibilité de la differenciation sexuelle chez l'Amphibien Urodele, *Pleurodeles waltl* toutes les femelles génétiques sous l'action de la temperature d'élevage. *Reproduction et Nutrition Developpement* 25, 671–688.

Downie, J.R. (1989) Observations on foam-making by *Leptodactylus fuscus* tadpoles. *Herpetological Journal* 1, 351–355.

Duellman, W.E. (1978) The biology of an equatorial herpetofauna in Amazonian Ecuador. *Miscellaneous Publications of the Museum of Natural History of the University of Kansas* 65, 1–352.

Duellman, W.E. and Trueb, L. (1986) *Biology of Amphibians*. McGraw-Hill Publishing Co., New York.

Dunson, W.A. (1977) Tolerance to high temperature and salinity by tadpoles of the Philippine frog, *Rana cancrivora*. *Copeia* 1977, 375–378.

Ebendal, T. and Uzzell, T. (1982) Ploidy and immunological distance in Swedish water frogs (*Rana esculenta* complex). *Amphibia–Reptilia* 3, 125–133.

Ehrenfeld, J.G. (1983) The effects of changes in land use on swamps of the New Jersey Pine Barrens. *Biological Conservation* 25, 353–375.

Elmberg, J. (1989) Knee-tagging: a new marking technique for anurans. *Amphibia–Reptilia* 10, 101–104.

Falkus, H. (1977) *Sea Trout Fishing* (2nd edn). H., F. and G. Witherby Ltd, London.

Fasola, M., Barbieri, F. and Canova, L. (1993) Test of an electronic individual tag for newts. *Herpetological Journal* 3, 149–150.

Ferguson, D.E. and Gilbert, C.C. (1967) Tolerances of three species of anuran amphibians to five chlorinated hydrocarbon insecticides. *Journal of the Mississippi Academy of Sciences* 13, 135–138.

Flower, R.J. and Battarbee, R.W. (1983) Diatom evidence for recent acidification of two Scottish lochs. *Nature* 305, 130–133.

Fog, K. (1988) The causes of decline of *Hyla arborea* on Bornholm. *Memoranda Soc. Fauna Flora Fennica* 64, 136–138.

Formanowicz, D.R., Stewart, M.M., Townsend, K. *et al.* (1981) Predation by giant crab spiders on the Puerto Rican frog *Eleutherodactylus coqui*. *Herpetologica* 37, 125–129.

Frazer, J.F.D. (1983) *Reptiles and Amphibians in Britain*. Collins, London.

Freda, J. and Dunson, W.A. (1984) Sodium balance of amphibian larvae exposed to low environmental pH. *Physiological Zoology* 57, 435–443.

Freda, J. and Dunson, W.A. (1985) The influence of external cation concentration on the hatching of amphibian embryos in water of low pH. *Canadian Journal of Zoology* 63, 2649–2656.

Freed, P.S. and Neitman, K. (1988) Notes on predation of the endangered Houston toad, *Bufo houstonensis*. *Texas Journal of Science* 40, 454–456.

Fuller, R.M. and Boorman, L.A. (1977) The spread and development of *Rhododendron ponticum* L. on dunes at Winterton, Norfolk, in comparison with invasion by *Hippophae rhamnoides* L. at Saltfleetby, Lincolnshire. *Biological Conservation* 12, 83–94.

Gabay, J.E. (1994) Ubiquitous natural antibiotics. *Science* 264, 373–374.

Gerhardt, H.C. (1975) Sound pressure levels and radiation patterns of the vocalisations of some North American frogs and toads. *Journal of Comparative Physiology* 102, 1–12.

Gill, D.E. (1978) The metapopulation ecology of the red-spotted newt, *Notophthalmus viridiscens* (Rafinesque). *Ecological Monographs* 48, 145–166.

Gill, D.E. (1979) Density dependence and homing behaviour in adult red-spotted newts *Notophthalmus viridiscens* (Rafinesque). *Ecology* 60, 800–813.

Gittins, S.P. (1983) Population dynamics of the common toad (*Bufo bufo*) at a lake in mid-Wales. *Journal of Animal Ecology* 52, 981–988.

Gittins, S.P., Kennedy, R.I. and Williams, R. (1984) Fecundity of the common toad (*Bufo bufo*) at a lake in mid-Wales. *British Journal of Herpetology* 6, 378–382.

Glandt, D. (1980) Die quantitative vertikalverbreitung der molch-arten, gattung *Triturus* (Amphibia, Urodela), in der Bundersreplublok Deutschland. *Bonn. Zool. Beitr.* 31, 97–110.

Glandt, D. (1986) Die saisonalen Wanderungen der mitteleuropaischen Amphibien. *Bonn Zool. Beitr.* 37, 211–228.

Goater, C.P. (1994) Growth and survival of postmetamorphic toads: interactions among larval history, density, and parasitism. *Ecology* 75, 2264–74.

Goater, C.P. and Ward, P.I. (1992) Negative effects of *Rhabdias bufonis* (Nematoda) on the growth and survival of toads (*Bufo bufo*). *Oecologia* 89, 161–165.

Golay, N. and Durrer, H. (1994) Inflammation due to toe-clipping in natterjack toads (*Bufo calamita*). *Amphibia–Reptilia* 15, 81–83.

Gosner, K.L. (1960) A simplified table for staging anuran embryos and larvae with notes on identification. *Herpetologica* 16, 183–190.

Graf, J-D. and Polls Pelaz, M. (1989) Evolutionary genetics of the *Rana esculenta* complex, in *Evolution and Ecology of Unisexual Vertebrates* (eds R.M. Dawley and J.P. Bogart). New York State Museum.

Grafen, A. (1990) Do animals really recognize kin? *Animal Behaviour* 39, 42–54.

Grant, D., Anderson, O. and Twitty, V.C. (1968) Homing orientation by olfaction in newts (*Taricha rivularis*). *Science* 160, 1354–1355.

Grayson, R.F. (1993) The distribution and conservation of the ponds of north-west England. *Lancashire Wildlife Journal* 2/3, 23–51.

Green, A.J. (1989) The sexual behaviour of the great crested newt, *Triturus cristatus* (Amphibia: Salamandridae). *Ethology* 83, 129–153.

Green, A.J. (1991) Large male crests, an honest indicator of condition, are preferred by female smooth newts *Triturus vulgaris* (Salamandridae) at the spermatophore transfer stage. *Animal Behaviour* 41, 367–369.

Green, D.M. (1983) Allozyme variation through a clinal hybrid zone between the toads *Bufo americanus* and *B. hemiophrys* in southeastern Manitoba. *Herpetologica* 39, 28–40.

Green, D.M. and Sessions, S.K. (1991) *Amphibian Cytogenetics and Evolution.* Academic Press, London.

Griffith, B., Scott, M.J., Carpenter, J.W. and Reed, C. (1989) Translocation as a species conservation tool: status and strategy. *Science* 245, 477–480.

Griffiths, R.A. (1985) A simple funnel trap for studying newt populations and an evaluation of trap behaviour in smooth and palmate newts *Triturus vulgaris* and *T. helveticus*. *Herpetological Journal* 1, 5–9.

Griffiths, R.A. (1986) Feeding niche overlap and food selection in smooth and palmate newts, *Triturus vulgaris* and *T. helveticus*, at a pond in mid-Wales. *Journal of Animal Ecology* 55, 201–214.

Griffiths, R.A. (1987) Microhabitat and seasonal niche dynamics of smooth and palmate newts, *Triturus vulgaris* and *Triturus helveticus*, at a pond in mid-Wales. *Journal of Animal Ecology* 56, 441–451.

Griffiths, R.A. (1991) Competition between common frog, *Rana temporaria*, and natterjack toad, *Bufo calamita*, tadpoles: the effect of competitor density and interaction level on tadpole development. *Oikos* 61, 187–196.

Griffiths, R.A. and Beebee, T.J.C. (1992) Decline and fall of the amphibians. *New Scientist* 134 (1827), 25–29.

Griffiths, R.A. and Denton, J. (1992) Interspecific associations in tadpoles. *Animal Behaviour* 44, 1153–1157.

Griffiths, R.A. and Mylotte, V.J. (1987) Microhabitat selection and feeding relations of smooth and warty newts, *Triturus vulgaris* and *T. cristatus*, at an upland pond in mid-Wales. *Holarctic Ecology* 10, 1–7.

Griffiths, R.A., Denton, J. and Wong, A.L.-C. (1993) The effect of food level on competition in tadpoles: interference mediated by protothecan algae? *Journal of Animal Ecology* 62, 274–279.

Griffiths, R.A., Edgar, P.W. and Wong, A.L-C. (1991) Interspecific competition in tadpoles: growth inhibition and growth retrieval in natterjack toads, *Bufo calamita*. *Journal of Animal Ecology* 60, 1065–1076.

Griffiths, R.A., Getliff, J.M. and Mylotte, V.J. (1988) Diel patterns of activity and vertical migration in tadpoles of the common toad, *Bufo bufo*. *Herpetological Journal* 1, 223–226.

Griffiths, R.A., Roberts, J.M. and Sims, S. (1987) A natural hybrid newt, *Triturus helveticus* × *T. vulgaris*, from a pond in mid-Wales. *Journal of Zoology (London)* 213, 133–140.

Griffiths, R.A., de Wijer, P. and Brady, L. (1993) The effect of pH on embryonic and larval development in smooth and palmate newts, *Triturus vulgaris* and *T. helveticus*. *Journal of Zoology (London)* 230, 401–409.

Griffiths, R.A., de Wijer, P. and May, R.T. (1994) Predation and competition within an assemblage of larval newts (*Triturus*). *Ecography* 17, 176–181.

Groombridge, B. (1988) *World Checklist of Threatened Amphibians and Reptiles*. Nature Conservancy Council, UK.

Grossenbacher, K. (1988) *Verbreitungsatlas der amphibien der Schweiz*. Dokumenta Faunistica Helvetiae. Schweiz Bund für Naturschutz, Basel.

Grubb, J.C. (1973) Olfactory orientation in breeding Mexican toads, *Bufo valliceps*. *Copeia* 1973, 490–496.

Gurdon, J.B. (1974) *The Control of Gene Expression in Animal Development*. Clarendon Press, Oxford.

Gutzke, W.H.N. (1987) Sex determination and sexual differentiation in reptiles. *Herpetological Journal* 1, 122–125.

Hagstrom, T. (1973) Identification of newt specimens (*Urodela, Triturus*) by recording the belly pattern and a description of photographic equipment for such registrations. *British Journal of Herpetology* 4, 321–326.

Hagstrom, T. (1980) Reproductive strategy and success of amphibians in waters acidified by atmospheric pollution. *Proceedings of the European Herpetological Symposium* 1980, 55–57.

Hairston, N.G. (1980) The experimental test of an analysis of field distributions: competition in terrestrial salamanders. *Ecology* 61, 817–826.

Hairston, N.G. (1983) Growth, survival and reproduction of *Plethodon jordani*: trade offs between selection pressures. *Copeia* 1983, 1024–1035.

Hairston, N.G. (1987) *Community Ecology and Salamander Guilds*. Cambridge University Press, Cambridge.

Halfpenny, G. (1978) *Atlas of the Amphibians and Reptiles of Staffordshire*. City Museum and Art Gallery, Stoke-on-Trent.

Hall, R.J. and Henry, P.F.P. (1992) Assessing effects of pesticides on amphibians and reptiles: status and needs. *Herpetological Journal* 2, 65–71.

Halliday, T.R. (1975) An observational and experimental study of sexual behaviour in the smooth newt, *Triturus vulgaris* (Amphibia: Salamandridae). *Animal Behaviour* 23, 291–322.

Halliday, T.R. and Adler, K. (1986) *The Encyclopaedia of Reptiles and Amphibians*. Equinox, Oxford.

Halliday, T.R. and Arano, B. (1991) Resolving the phylogeny of the European newts. *Trends in Ecology and Evolution* 6, 113–117.

Halliday, T.R. and Verrell, P.A. (1986) Sexual selection and body size in amphibians. *Herpetological Journal* 1, 86–92.

Halliday, T.R. and Verrell, P.A. (1988) Body size and age in amphibians and reptiles. *Journal of Herpetology* 22, 253–265.

Harrison, J.D. (1987) Food and feeding relations of common frog and common toad tadpoles (*Rana temporaria* and *Bufo bufo*) at a pond in mid-Wales. *Herpetological Journal* 1, 141–143.

Harrison, J.D., Gittins, S.P. and Slater, F.M. (1983) The breeding migrations of smooth and palmate newts (*Triturus vulgaris* and *T. helveticus*) at a pond in mid-Wales. *Journal of Zoology (London)* 199, 249–258.

Harrison, R.G. (1969) Harrison stages and description of the normal development of the spotted salamander, *Ambystoma punctatum* (Lin.), in *Organization and Development of the Embryo* (ed. R.G. Harrison). Yale University Press, New Haven.

Hazelwood, E. (1969) A study of a breeding colony of frogs at the Canon Slade Grammar School, near Bolton, Lancs. *British Journal of Herpetology* 4, 96–102.

Hazelwood, E. (1970) Frog pond contaminated. *British Journal of Herpetology* 4, 177–185.

Hedlund, L. (1990) Factors affecting differential mating success in male crested newts, *Triturus cristatus*. *Journal of Zoology (London)* 220, 33–40.

Heinen, J.T. (1993) Aggregations of newly metamorphosed *Bufo americanus*: tests of two hypotheses. *Canadian Journal of Zoology* 71, 334–338.

Heinzmann, U. (1970) Untersuchen zur Bio-Akustik und Okologie der Geburtshelferkrote, *Alytes o. obstetricans* (Laur). *Oecologia* 5, 19–55.

Hemelaar, A.S.M. (1981) Age determination of male *Bufo bufo* (Amphibia, Anura) from the Netherlands, based on year rings in phalanges. *Amphibia–Reptilia* 1, 223–233.

Hemelaar, A.S.M. (1983) Age of *Bufo bufo* in amplexus over the spawning period. *Oikos* 40, 1–5.

Hemelaar, A.S.M. (1986) Demographic study on *Bufo bufo* L. (Anura, Amphibia) from different climates, by means of skeletochronology. PhD Thesis, University of Nijmegen.

Hemmer, H. (1973) Die bastardierung von kreutzkrote (*Bufo calamita*) und wech-selkrote (*Bufo viridis*). *Salamandra* 9, 118–136.

Hemmer, H., Kadel, B. and Kadel, K. (1981) The Balearic toad (*Bufo viridis balearicus* (BOETTGER, 1881)), human bronze age culture, and Mediterranean biogeography. *Amphibia–Reptilia* 2, 217–230.

Hensley, F.R. (1993) Ontogenetic loss of phenotypic plasticity of age at metamorphosis in tadpoles. *Ecology* 74, 2405–2412.

Hews, D.K. and Blaustein, A.R. (1985) An investigation of the alarm response in *Bufo boreas* and *Rana cascadae* tadpoles. *Behavioural and Neural Biology* 43, 47–57.

Heyer, W.R. (1973) Ecological interactions of frog larvae at a seasonal tropical location in Thailand. *Journal of Herpetology* 7, 337–361.

Heyer, W.R. (1976) Studies in larval habitat partitioning. *Smithsonian Contributions to Zoology* 242, 1–27.

Heyer, W.R, Rand, A.S., Goncalvez da Cruz, C.A. and Peixoto, O.L. (1988) Decimation, extinction and colonisation of frog populations in southeast Brazil and their evolutionary implications. *Biotropica* 20, 230–235.

Hillis, D. and Moritz, C. (1990) *Molecular Systematics.* Sinauer, Sunderland, Mass.

Hillis, D.M., Hillis, A.M. and Martin, R.F. (1984) Reproductive ecology and hybridisation of the endangered Houston toad (*Bufo houstonensis*). *Journal of Herpetology* 18, 56–72.

Hodl, W. and Gollmann, G. (1986) Distress calls in neotropical frogs. *Amphibia–Reptilia* 7, 11–21.

Hoelzel, A.R. (1992) *Molecular Genetic Analysis of Populations.* IRL Press, Oxford.

Hoffman, J. and Katz, U. (1989) The ecological significance of burrowing behaviour in the toad (*Bufo viridis*). *Oecologia* 81, 510–513.

Hoffmann, A.A. and Blows, M.W. (1994) Species borders: ecological and evolutionary perspectives. *Trends in Ecology and Evolution* 9, 223–227.

Hoglund, J. and Robertson, J.G.M. (1987) Random mating by size in a population of common toads (*Bufo bufo*). *Amphibia–Reptilia* 8, 321–330.

Holman, J.A. (1993) British quaternary herpetofaunas: a history of adaptations to Pleistocene disruptions. *Herpetological Journal* 3, 1–7.

Holman, J.A. and Stuart, A.J. (1991) Amphibians of the Whitemoor channel early Flandrian site near Bosley, east Cheshire; with remarks on the fossil distribution of *Bufo calamita* in Britain. *Herpetological Journal* 1, 568–573.

Holomuzki, J.R., Collins, J.P. and Brunkow, P.E. (1994) Trophic control of fishless ponds by tiger salamander larvae. *Oikos* 71, 55–64.

Honegger, R.E. (1981) List of amphibians and reptiles either known or thought to have become extinct since 1600. *Biological Conservation* 19, 141–158.

Hoperskaya, O.A. (1975) The development of an animal homozygous for a mutation causing periodic albinism (a^p) in *Xenopus laevis*. *Journal of Embryology and Experimental Morphology* 34, 253–264.

Hotz, H., Beerli, P. and Spolsky, C. (1992) Mitochondrial DNA reveals formation of nonhybrid frogs by natural matings between hemiclonal hybrids. *Molecular Biology and Evolution* 9, 610–620.

Houck, L.D. (1988) The effect of body size on male courtship success in a plethodontid salamander. *Animal Behaviour* 36, 837–842.

Howard, R.D. (1978) The evolution of mating strategies in bullfrogs, *Rana catesbiana*. *Evolution* 32, 850–871.

Hurlbert, S.H. (1978) The measurement of niche overlap and some relatives. *Ecology* 59, 67–77.

Hutchison, M.N. and Maxson, L.R. (1987) Biochemical studies on the relationships of the gastric-brooding frogs, genus *Rheobatrachus*. *Amphibia–Reptilia* 8, 1–12.

Inger, R.F., Shaffer, H.B., Koshy, M. and Bakde, R. (1987) Ecological structure of a herpetological assemblage in South India. *Amphibia–Reptilia* 8, 189–202.

Ishchenko, V.G. (1989) Population biology of amphibians. *Soviet Science Reviews for Physiology and General Biology,* 3, 119–155.

Jacobson, N.L. (1989) Breeding dynamics of the Houston toad. *Southwestern Naturalist* 34, 374–380.

Jaeger, R.G. (1971a) Competitive exclusion as a factor influencing the distribution of two species of terrestrial salamanders. *Ecology* 52, 632–637.

Jaeger, R.G. (1971b) Moisture as a factor influencing the distributions of two species of terrestrial salamanders. *Oecologia* 6, 191–207.

Jaeger, R.G. (1981) Dear enemy recognition and the costs of aggression between salamanders. *American Naturalist* 117, 962–974.

Joensen, A.H. (1967) Urfuglen (*Lyurus tetralix*): Denmark. *Danske VildfundersØgelser* 14, 1–102.

Johnson, R.R. (1992) Habitat loss and declining amphibian populations. *Canadian Wildlife Service Occasional Papers* 76, 71–75.

Kadel, K. (1975) Freilandstudien zur Uberlebensrate von Kreuzkrotenlarven (*Bufo calamita* LAUR.). *Revue suisse Zoologische* 82, 237–244.

Kalezic, M.L. (1984) Evolutionary divergences in the smooth newt, *Triturus vulgaris* (Urodela, salamandridae): Electrophoretic evidence. *Amphibia–Reptilia* 5, 221–230.

Kats, L.B., Petranka, J.W. and Sih, A. (1988) Antipredator defenses and the persistence of amphibian larvae with fishes. *Ecology* 69, 1865–1870.

Keller, L.F., Arcese, P., Smith, J.N.M. *et al.* (1994) Selection against inbred song sparrows during a natural population bottleneck. *Nature* 372, 356–7.

Kluge, A.G. (1981) The life history, social organisation and parental behaviour of *Hyla rosenbergi* Boulenger, a nest-building gladiator frog. *Miscellaneous Publications of the Museum of Zoology of the University of Michigan* 160, 1–170.

Kottmann, H.J., Schwoppe, W., Willers, T. and Wittig, R. (1985) Heath conservation by sheep grazing: a cost–benefit analysis. *Biological Conservation* 31, 67–74.

Kruse, K.C. (1983) Optimal foraging by predaceous diving beetle larvae on toad tadpoles. *Oecologia* 58, 383–388.

Kuhn, J. (1993) Fortpflanzungsbiologie der Erdkrote *Bufo b. bufo* (L.) in einer Wildflubaue. *Zeitschrift für Okologie und Naturschutz* 2, 1–10.

Kuzmin, S.L. (1991) Food resource allocation in larval newt guilds (genus *Triturus*). *Amphibia–Reptilia* 12, 293–304.

Laan, R. and Verboom, B. (1990) Effects of pool size and isolation on amphibian communities. *Biological Conservation* 54, 251–262.

Langton, T.E.S. (1989) *Amphibians and Roads*. ACO Polymer Products, Shefford, England.

Langton, T.E.S. (1991) Distribution and status of reptiles and amphibians in the London area. *London Naturalist* 70, 97–123.

Larsen, L.O. and Pedersen, J.N. (1982) The snapping response of the toad, *Bufo bufo*, towards prey dummies at very low light intensities. *Amphibia–Reptilia* 2, 321–328.

Lawler, S.P. (1989) Behavioural responses to predators and predation risk in four species of larval anurans. *Animal Behaviour* 38, 1039–1047.

Lawler, S.P. and Morin, P.J. (1993) Temporal overlap, competition, and priority effects in larval anurans. *Ecology* 74, 174–182.

Leips, J. and Travis, J. (1994) Metamorphic responses to changing food levels in two species of hylid frogs. *Ecology* 75, 1345–1356.

Levins, R. (1968) *Evolution in Changing Environments*. Princeton University Press.

Lizana, M., Perez-Mellado, V. and Ciudad, M.J. (1990) Analysis of the structure of an amphibian community in the central system of Spain. *Herpetological Journal* 1, 435–446.

Loftus-Hills, J.J. and Littlejohn, M.J. (1971) Mating-call sound intensities of anuran amphibians. *Journal of the Acoustical Society of America* 49, 1327–1329.

Macgregor, H.C. (1978) Some trends in the evolution of very large chromosomes. *Philosophical Transactions of the Royal Society of London* B, 283, 309–318.

Macgregor, H.C. and Horner, H. (1980) Heteromorphism for chromosome 1, a requirement for normal development in crested newts. *Chromosoma* 76, 111–122.

Macgregor, H.C. and Varley, J. (1983) *Working with Animal Chromosomes*. John Wiley and Sons, Chichester.

Macgregor, H.C., Sessions, S.K. and Arntzen, J.W. (1990) An integrative analysis of phylogenetic relationships among newts of the genus *Triturus* (family Salamandridae), using comparative biochemistry, cytogenetics and reproductive interactions. *Journal of Evolutionary Biology* 3, 329–373.

Madler, H. (1984) Animal habitat isolation by roads and agricultural fields. *Biological Conservation* 29, 81–96.

Marrs, R.H. (1984) Birch control on lowland heaths: mechanical control and the application of selective herbicides by foliar spray. *Journal of Applied Ecology* 21, 703–716.

Marrs, R.H. (1987) Studies on the conservation of lowland Calluna heaths: 1. Control of birch and bracken and its effect on heath vegetation. *Journal of Applied Ecology* 24, 163–175.

Marrs, R.H., Hicks, M.J. and Fuller, R.M. (1986) Losses of lowland heath through succession at four sites in Breckland, East Anglia, England. *Biological Conservation* 36, 19–38.

Martins, M. (1993) Observations on the reproductive behaviour of the Smith frog *Hyla faber*. *Herpetological Journal* 3, 31–34.

Mattison, C. (1987) *Frogs and Toads of the World*. Blandford Press, Poole, Dorset.

Maxson, L.R. (1984) Molecular probes of phylogeny and biogeography in toads of the widespread genus *Bufo*. *Molecular Biology and Evolution* 1, 345–356.

Mayol, J. and Alcover, J.A. (1981) Survival of *Baleaphryne* SANCHIZ and ANDROVER, 1977 (Amphibia: Anura: Discoglossidae) on Mallorca. *Amphibia–Reptilia* 1, 343–345.

McCormick, S. and Polis, G.A. (1982) Arthropods that prey on vertebrates. *Biological Reviews* 57, 29–58.

McMillan, N.F. (1963) Toads continue to migrate for spawning to a now vanished pond. *British Journal of Herpetology* 3, 88.

Miaud, C. (1993) Predation on newt eggs (*Triturus alpestris* and *T. helveticus*): identification of predators and protective role of oviposition behaviour. *Journal of Zoology (London)* 231, 575–582.

Miaud, C., Joly, P. and Castanet, J. (1993) Variation in age structures in a subdivided population of *Triturus cristatus*. *Canadian Journal of Zoology* 71, 1874–1879.

Montori, A. (1989) Skeletochronological results in the Pyrenean newt *Euproctus asper* (Duges, 1852) in two pre-Pyrenean populations. *Abstracts of the First World Congress of Herpetology*, University of Kent.

Morin, P.J. (1983) Predation, competition, and the composition of larval anuran guilds. *Ecological Monographs* 53, 119–138.

Morin, P.J. and Johnson, E.A. (1988) Experimental studies of asymmetric competition among anurans. *Oikos* 53, 398–407.

Morin, P.J., Lawler, S.P. and Johnson, E.A. (1988) Competition between aquatic insects and vertebrates: interaction strength and higher order interactions. *Ecology* 69, 1401–1409.

Morin, P.J., Lawler, S.P. and Johnson, E.A. (1990) Ecology and breeding phenology of larval *Hyla andersonii*: the disadvantages of breeding late. *Ecology* 71, 1590–1598.

Moritz, C. (1994) Applications of mitochondrial DNA analysis in conservation: a critical review. *Molecular Ecology* 3, 401–411.

Nei, M. (1987) *Molecular Evolutionary Genetics*. Columbia University Press, New York.

Nevo, E. and Beiles, A. (1991) Genetic diversity and ecological heterogeneity in amphibian evolution. *Copeia* 1991, 565–597.

Nevo, E., Beiles, A. and Ben-Schlomo, R. (1984) The evolutionary significance of genetic diversity: ecological, demographic and life-history correlates, in *Evolutionary Dynamics of Genetic Diversity* (ed. G.S. Mani). Springer, Berlin.

Newman, R.A. (1989) Developmental plasticity of *Scaphiopus couchii* tadpoles in an unpredictable environment. *Ecology* 70, 1775–1787.

Nichols, O.G. and Bamford, M.J. (1985) Reptile and frog utilisation of rehabilitated bauxite minesites and dieback-affected sites in Western Australia's Jarrah *Eucalyptus marginata* forest. *Biological Conservation* 34, 227–249.

Nishioka, M., Sumida, M., Ueda, H. and Wu, Z. (1990) Genetic relationships among 13 *Bufo* species and subspecies elucidated by the method of electrophoretic analyses. *Scientific Reports of the Laboratory of Amphibian Biology (Hiroshima University)* 10, 53–91.

Nunney, L. and Campbell, K.A. (1993) Assessing minimum viable population size: demography meets population genetics. *Trends in Ecology and Evolution* 8, 234–239.

Oldham, R.S. and Nicholson, M. (1986) *Status and Ecology of the Warty Newt, Triturus cristatus.* Nature Conservancy Council Report, Peterborough, Cambs.

Oldham, R.S. and Swan, M.J.S. (1991) Conservation of amphibian populations in Britain, in *Species Conservation: A population-biological approach* (eds A. Seitz and V. Loeschcke), Birkhäuser Verlag, Basel.

Oldham, R.S. and Swan, M. (1993) Pond loss – the present position, in *Proceedings of the conference protecting Britain's ponds* (eds C. Aistrop and J. Biggs). Wildfowl & Wetlands Trust and Pond Action, Oxford.

Olmo, E. (1991) Genome variations in the transition from Amphibians to Reptiles. *Journal of Molecular Evolution* 33, 68–75.

Olson, D.H., Blaustein, A.R. and O'Hara, R.K. (1986) Mating pattern variability among western toad (*Bufo boreas*) populations. *Oecologia* 70, 351–356.

Ormerod, E.A. (1872) Observations on the cutaneous exudation of the *Triturus cristatus* or great water newt. *Journal of the Linnean Society* 11, 493–496.

Passmore, N.I., Bishop, P.J. and Caithness, N. (1992) Calling behaviour influences mating success in male painted reed frogs, *Hyperolius marmoratus*. *Ethology* 91, 237–247.

Pavignano, I. (1989) Method employed to study the diet of anuran amphibian larvae. *Amphibia–Reptilia* 10, 453–456.

Pavignano, I., Giacoma, C. and Castellano, S. (1990) A multivariate analysis of amphibian habitat determinants in north western Italy. *Amphibia–Reptilia* 11, 311–324.

Pearman, P.B. (1993) Effects of habitat size on tadpole populations. *Ecology* 74, 1982–1991.

Pechmann, J.H.K. and Wilbur, H.M. (1994) Putting declining amphibian populations in perspective: natural fluctuations and human impacts. *Herpetologica* 50, 65–84.

Pechmann, J.H.K., Scott, D.E., Semlitsch, R.D. *et al.* (1991) Declining amphibian populations: the problem of separating human impacts from natural fluctuations. *Science* 253, 892–895.

Pecio, A. (1992) Insemination and egg-laying dynamics in the smooth newt, *Triturus vulgaris*, in the laboratory. *Herpetological Journal* 2, 5–7.

Peterson, J.A. and Blaustein, A.R. (1991) Unpalatability in anuran larvae as a defense against natural salamander predators. *Ethology, Ecology and Evolution* 3, 63–72.

Petranka, J.W. (1989) Chemical interference competition in tadpoles: does it occur outside laboratory aquaria? *Copeia* 1989, 921–930.

Pfennig, D.W., Reeve, H.K. and Sherman, P.W. (1993) Kin recognition and cannibalism in spadefoot toad tadpoles. *Animal Behaviour* 46, 87–94.

Phillips, J.B. (1986) Magnetic compass orientation in the eastern red-spotted newt *Notophthalmus viridiscens. Journal of Comparative Physiology (A)* 158, 103–109.

Pianka, E.R. (1970) On 'r' and 'K' selection. *American Naturalist* 104, 592–597.

Picker, M.D. and deVilliers, A.L. (1989) The distribution and conservation status of *Xenopus gilli* (Anura: pipidae). *Biological Conservation* 49, 169–183.

Plytycz, B. and Bigaj, J. (1993) Studies on the growth and longevity of the yellow-bellied toad, *Bombina variegata*, in natural environments. *Amphibia–Reptilia* 14, 35–44.

Polls Pelaz, M. (1990) The biological klepton concept (BKC). *Alytes* 8, 78–89.

Pough, F.H. (1976) Acid precipitation and embryonic mortality of spotted salamanders, *Ambystoma maculatum. Science* 192, 68–70.

Pound, J.A. and Crump, M.L. (1994) Amphibian declines and climate disturbance: the case of the golden toad and the harlequin frog. *Conservation Biology* 8, 72–85.

Power, T., Clark, K.L., Harfenist, A. and Peakall, D.B. (1989) A review and evaluation of the amphibian toxicological literature. *Canadian Wildlife Services Technical Report* Series 61.

Prendergast, J.R., Quinn, R.M., Lawton, J.H. *et al.* (1993) Rare species, the coincidence of diversity hotspots and conservation strategies. *Nature* 365, 335–337.

Prestt, I., Cooke, A.S. and Corbett, K.F. (1974) British amphibians and reptiles, in *The Changing Flora and Fauna of Britain* (ed. D.L. Hawksworth), pp. 229–254. Academic Press, London.

Raup, D.M. (1991) *Extinction: Bad Genes or Bad Luck?* W.W. Norton, New York.

Raxworthy, C.J. (1990) A review of the smooth newt (*Triturus vulgaris*) subspecies, including an identification key. *Herpetological Journal* 1, 481–492.

Raxworthy, C.J., Kjolbye-Biddle, B. and Biddle, M. (1990) An archaeological study of frogs and toads from the eighth to the sixteenth century at Repton, Derbyshire. *Herpetological Journal* 1, 504–509.

Reading, C.J. (1984) Interspecific spawning between common frogs (*Rana temporaria*) and common toads (*Bufo bufo*). *Journal of Zoology (London)* 203, 95–101.

Reading, C.J. and Clarke, R.T. (1983) Male breeding behaviour and mate acquisition in the common toad *Bufo bufo. Journal of Zoology (London)* 201, 237–246.

Reading, C.J., Loman, J. and Madsen, T. (1991) Breeding pond fidelity in the common toad, *Bufo bufo. Journal of Zoology (London)* 225, 201–211.

Reh, W. and Seitz, A. (1990) The influence of land use on the genetic structure of populations of the common frog *Rana temporaria. Biological Conservation* 54, 239–249.

Richards, C.M. (1958) The inhibition of growth in crowded *Rana pipiens* tadpoles. *Physiological Zoology* 31, 138–151.

Richards, C.M. (1962) The control of tadpole growth by alga-like cells. *Physiological Zoology* 35, 285–296.

Riis, N. (1991) A field study of survival, growth, biomass and temperature dependence of *Rana dalmatina* and *Rana temporaria* larvae. *Amphibia–Reptilia* 12, 229–244.

Roithmuir, M.E. (1992) Territoriality and male mating success in the dart-poison frog, *Epipedobates femoralis* (Dendrobatidae, anura). *Ethology* 92, 331–343.

Rose, S.M. (1960) A feedback mechanism of growth control in tadpoles. *Ecology* 41, 188–199.

Rowe, C.L. and Dunson, W.A. (1994) The value of simulated pond communities in mesocosms for studies of amphibian ecology and ecotoxicology. *Journal of Herpetology* 28, 346–356.

Rowell, T.A. (1986) The history of drainage at Wicken Fen, Cambridgeshire, England and its relevance to conservation. *Biological Conservation* 35, 111–142.

Rowell, T.A. (1991) *SSSIs: A Health Check*. Wildlife Link, London.

Ryan, M.J. (1980) Female mate choice in a Neotropical frog. *Science* 209, 523–525.

Ryder, O.A. (1986) Species conservation and systematics: the dilemma of subspecies. *Trends in Ecology and Evolution* 1, 9–10.

Ryser, J. (1989) Weight loss, reproductive output and the cost of reproduction in the common frog *Rana temporaria*. *Oecologia* 78, 264–268.

Salvidio, S., Lattes, A., Tavano, M. et al. (1994) Ecology of a *Speleomantes ambrosii* population inhabiting an artificial tunnel. *Amphibia–Reptilia* 15, 35–46.

Samollow, P.B. (1980) Selective mortality and reproduction in a natural population of *Bufo boreas*. *Evolution* 34, 18–39.

Savage, R.M. (1961) *The Ecology and Life History of the Common Frog Rana temporaria temporaria*. Pitman, London.

Schindler, D.W., Mills, K.H., Malley, D.F. et al. (1985) Long term ecosystem stress: the effects of years of experimental acidification on a small lake. *Science* 228, 1395–1401.

Schlupp, I. and Podloucky, R. (1994) Changes in breeding site fidelity: a combined study of conservation and behaviour in the common toad *Bufo bufo*. *Biological Conservation* 69, 285–291.

Schlyter, F., Hoglund, J. and Stromberg, G. (1991) Hybridisation and low numbers in isolated populations of the natterjack, *Bufo calamita*, and the green toad, *B. viridis*, in southern Sweden: possible conservation problems. *Amphibia–Reptilia* 12, 267–281.

Schmid, W.D. (1982) Survival of frogs in low temperature. *Science* 215, 697–698.

Schmidt, B.R. (1993) Are hybridogenetic frogs cyclical parthenogens? *Trends in Ecology and Evolution* 8, 271–273.

Schoener, T.W. (1970) Nonsynchronous spatial overlap of lizards in patchy habitats. *Ecology* 51, 408–418.

Schoorl, J. and Zuiderwijk, A. (1981) Ecological isolation in *Triturus cristatus* and *Triturus marmoratus* (Amphibia: Salamandridae). *Amphibia–Reptilia* 3/4, 235–252.

Scott, D.E. (1994) The effect of larval density on adult demographic traits in *Ambystoma opacum*. *Ecology* 75, 1383–1396.

Scribner, K.T., Arntzen, J.W. and Burke, T. (1994) Comparative analysis of intra- and interpopulation genetic diversity in *Bufo bufo*, using allozyme, single-locus microsatellite, minisatellite and multilocus minisatellite data. *Molecular Biology and Evolution* 11, 737–748.

Seale, D.B. (1980) Influence of amphibian larvae on primary production, nutrient flux and competition in a pond ecosystem. *Ecology* 61, 1531–1550.

Semlitsch, R.D. (1993) Effects of different predators on the survival and development of tadpoles from the hybridogenetic *Rana esculenta* complex. *Oikos* 67, 40–46.

Semlitsch, R.D. and Gavasso, S. (1992) Behavioural responses of *Bufo bufo* and *Bufo calamita* tadpoles to chemical cues of vertebrate and invertebrate predators. *Ethology, Ecology and Evolution* 4, 165–173.

Semlitsch, R.D. and Wilbur, H.M. (1989) Artificial selection for paedomorphosis in the salamander *Ambystoma talpoideum*. *Evolution* 43, 105–112.

Sexton, O.J., Phillips, C.A. and Routman, E. (1994) The response of naive breeding adults of the spotted salamander to fish. *Behaviour* 130, 113–121.

Shine, R. (1979) Sexual selection and sexual dimorphism in the Amphibia. *Copeia* 1979, 297–306.

Shine, R. (1994) National peculiarities, scientific traditions and research directions. *Trends in Ecology and Evolution* 9, 309.

Silverin, B. and Andren, C. (1992) The ovarian cycle in the natterjack toad, *Bufo calamita*, and its relation to breeding behaviour. *Amphibia–Reptilia* 13, 177–192.

Sinsch, U. (1987) Orientation behaviour of toads (*Bufo bufo*) displaced from the breeding site. *Journal of Comparative Physiology (A)* 161, 715–727.

Sinsch, U. (1988) Seasonal changes in the migratory behaviour of the toad *Bufo bufo*: direction and magnitude of movements. *Oecologia* 76, 390–398.

Sinsch, U. (1989) Die kreuzkrote (*Bufo calamita*): dynamik und mikrohabitate einer kiesgrubenpopulation. *Poster zu verhandlungen der Gesellschaft fur Okologie (Essen)* 18, 101–109.

Sinsch, U. (1990a) Migration and orientation in anuran amphibians. *Ethology, Ecology and Evolution* 2, 65–79.

Sinsch, U. (1990b) The orientation behaviour of three toad species (genus *Bufo*) displaced from the breeding site. *Fortschritte der Zoologie* 38, 75–83.

Sinsch, U. (1991) The orientation behaviour of amphibians. *Herpetological Journal* 1, 541–544.

Sinsch, U. (1992a) Sex-biased site fidelity and orientation behaviour in reproductive natterjack toads (*Bufo calamita*). *Ethology, Ecology and Evolution* 4, 15–32.

Sinsch, U. (1992b) Structure and dynamic of a natterjack toad metapopulation (*Bufo calamita*). *Oecologia* 90, 489–499.

Sjögren, P. (1991) Extinction and isolation gradients in metapopulations: the case of the pool frog (*Rana lessonae*). *Biological Journal of the Linnean Society* 42, 135–147.

Sjögren Gulve, P. (1994) Distribution and extinction patterns within a northern metapopulation of the pool frog, *Rana lessonae*. *Ecology* 75, 1357–1367.

Skelly, D.K. and Werner, E.E. (1990) Behavioural and life-historical responses of larval American toads to an odonate predator. *Ecology* 71, 2313–2322.

Smith, D.C. (1990) Population structure and competition among kin in the chorus frog (*Pseudacris triseriata*). *Evolution* 44, 1529–1541.

Smith, F.D.M., May, R.M., Pellew, R. *et al.* (1993) How much do we know about the current extinction rate? *Trends in Ecology and Evolution* 8, 375–378.

Smith, H.M. (1978) *Amphibians of North America*. Golden Press, New York.

Smith, M.A. (1964) *The British Amphibians and Reptiles* (3rd edn). Collins, London.

Smith, P.H. and Payne, K.R. (1980) A survey of natterjack toad *Bufo calamita* distribution and breeding success in the north Merseyside sand-dune system. *Biological Conservation* 19, 27–39.

Smith-Gill, S.J. and Berven, K.A. (1979) Predicting amphibian metamorphosis. *American Naturalist* 113, 563–585.

Sparreboom, M. and Teunis, B. (1990) The courtship display of the marbled newt, *Triturus m. marmoratus*. *Amphibia–Reptilia* 11, 351–361.

Springer, M.S., Davidson, E.H. and Britten, R.J. (1991) Calculation of sequence divergence from the thermal stability of DNA heteroduplexes. *Journal of Molecular Evolution* 34, 379–382.

Spurway, H. (1953) Genetics of specific and subspecific differences in European newts. *Symposium of the Society for Experimental Biology* 7, 200–238.

Sredl, M.J. and Collins, J.P. (1991) The effect of ontogeny on interspecific interactions in larval amphibians. *Ecology* 72, 2232–2239.

Steinwascher, K. (1979) Host–parasite interaction as a potential population-regulating mechanism. *Ecology* 60, 884–890.

Stenhouse, S.L. (1985a) Migration, orientation and homing in *Ambystoma maculatum* and *Ambystoma opacum*. *Copeia* 1985, 631–637.

Stenhouse, S.L. (1985b) Interdemic variation in predation on salamander larvae. *Ecology* 66, 1706–1717.

Stenhouse, S.L., Hairston, N.G. and Cobey, A.E. (1983) Predation and competition in *Ambystoma* larvae: field and laboratory experiments. *Journal of Herpetology* 17, 210–220.

Steward, J.W. (1969) *The Tailed Amphibians of Europe*. David & Charles, Newton Abbot, Devon.

Stocks, R. (1992) Mycophagy in a fossorial microhylid *Copiula fistulans* in New Guinea. *Herpetological Journal* 2, 61–63.

Strijbosch, H. (1979) Habitat selection of amphibians during their aquatic phase. *Oikos* 33, 363–372.

Strijbosch, H. (1980a) Habitat selection by amphibians during their terrestrial phase. *British Journal of Herpetology* 6, 93–98.

Strijbosch, H. (1980b) Mortality in a population of *Bufo bufo* resulting from the fly *Lucilia bufonivora*. *Oecologia* 45, 285–286.

Strijbosch, H. (1992) Niche segregation in the lizards and frogs of a lowland heath. *Proceedings of the 6th Ordinary General Meeting of the Societas Europaea Herpetologica*, 415–420.

Stumpel, A.H.P. and Tester, U. (1992) *Ecology and Conservation of the European Tree Frog*. DLO Institute for Forestry and Nature Research, Wageningen, Netherlands.

Summers, K. (1992) Dart-poison frogs and the control of sexual selection. *Ethology* 91, 89–107.

Swan, M.J.S. and Oldham, R.S. (1993) *Herptile Sites, Volume 1: National Amphibian Survey Final Report*. English Nature, Peterborough, Cambs.

Szymura, J.M. (1983) Genetic differentiation between hybridising species *Bombina bombina* and *Bombina variegata* (Salientia, Discoglossidae) in Poland. *Amphibia–Reptilia* 4, 137–146.

Szymura, J.M. and Barton, N.H. (1986) Genetic analysis of a hybrid zone between the fire-bellied toads, *Bombina bombina* and *B. variegata*, near Cracow in southern Poland. *Evolution* 40, 1141–1159.

Taylor, R.H.R. (1948) The distribution of reptiles and amphibia in the British Isles, with notes on species recently introduced. *British Journal of Herpetology* 1, 1–38.

Taylor, R.H.R. (1963) The distribution of amphibians and reptiles in England and Wales, Scotland and Ireland and the Channel Islands: a revised survey. *British Journal of Herpetology* 9, 95–116.

Tejedo, M. (1991) Effect of predation by two species of sympatric tadpoles on embryo survival in natterjack toads (*Bufo calamita*). *Herpetologica* 47, 322–327.

Tejedo, M. (1992) Large male mating advantage in natterjack toads, *Bufo calamita*: sexual selection or energetic constraints? *Animal Behaviour* 44, 557–569.

Tilley, S.G. and Hausman, J.S. (1976) Allozymic variation and occurrence of multiple inseminations in populations of the salamander *Desmognathus ochrophaeus*. *Copeia* 1976, 734–741.

Tilman, D., May, R.M., Lehman, C.L. and Nowak, M.A. (1994) Habitat destruction and the extinction debt. *Nature* 371, 65–66.

Tinsley, R.C. and Jackson, H.C. (1988) Pulsed transmission of *Pseudodiplorchis americanus* (Monogenea) between desert hosts (*Scaphiopus couchii*). *Parasitology* 97, 437–452.

Tocque, K. (1993) The relationship between parasite burden and host resources in the desert toad (*Scaphiopus couchii*), under natural environmental conditions. *Journal of Animal Ecology* 62, 683–693.

Toft, C.A. (1980) Feeding ecology of thirteen syntopic species of anurans in a seasonal tropical environment. *Oecologia* 45, 131–141.

Toft, C.A. (1985) Resource partitioning in amphibians and reptiles. *Copeia* 1985, 1–21.

Townsend, D.S., Stewart, M.M., Pough, F.H. and Brussard, P.F. (1981) Internal fertilisation in an oviparous frog. *Science* 212, 469–471.

Tracy, C.R. and Dole, J.W. (1969) Orientation of displaced California toads, *Bufo boreas*, to their breeding sites. *Copeia* 1969, 693–700.

Travis, J. (1984) Anuran size at metamorphosis: experimental test of a model based on intraspecific competition. *Ecology* 65, 1155–1160.

Travis, J., Emerson, S.B. and Blouin, M. (1987) A quantitative-genetic analysis of larval life-history traits in *Hyla crucifer*. *Evolution* 41, 145–156.

Tuttle, M.D. and Ryan, M.J. (1981) Bat predation and the evolution of frog vocalisation in the tropics. *Science* 214, 677–678.

Twitty, V.C. (1955) Field experiments on the biology and genetic relationships of the California species of *Triturus*. *Journal of Experimental Zoology* 129, 129–147.

Twitty, V.C., Grant, D. and Anderson, O. (1967) Long distance homing in the newt *Taricha rivularis*. *Proceedings of the National Acadamy of Sciences USA* 50, 51–58.

Tyler, M.J. (1976) *Frogs*. Collins, Sydney.

Tyler, M.J. and Davies, M. (1985) in *The Biology of Australasian Frogs and Reptiles* (eds G. Grigg, R. Shine and H. Ehmann). Surrey Beatty, Chipping Norton.

Tyler-Jones, R., Beattie, R.C. and Aston, R.J. (1989) The effects of acid water and aluminium on the embryonic development of the common frog, *Rana temporaria*. *Journal of Zoology (London)* 219, 355–372.

Van der Meulen, F. (1982) Vegetation changes and water catchment in a Dutch coastal dune area. *Biological Conservation* 24, 305–316.

Van Gelder, J.J., Aarts, H.M.J. and Staal, H.W.M. (1986) Routes and speed of migrating toads (*Bufo bufo* L.): a telemetric study. *Herpetological Journal* 1, 111–114.

Van Gelder, J.J., Olders, J.H.J., Bosch, J.W.G. and Starmans, P.W. (1986) Behaviour and body temperature of hibernating common toads *Bufo bufo*. *Holarctic Ecology* 9, 225–228.

Veith, M. (1987) Egg and embryo proteins in European newts (genus *Triturus*) and their taxonomic potential. *Amphibia–Reptilia* 8, 203–212.

Veith, M. (1992) The fire salamander, *Salamandra salamandra* L., in central Europe: subspecies distribution and intergradation. *Amphibia–Reptilia* 13, 297–314.

Vences, M. (1993) Habitat choice of the salamander *Chioglossa lusitanica*: the effects of eucalypt plantations. *Amphibia–Reptilia* 14, 201–212.

Verrell, P.A. (1984) The responses of inseminated female smooth newts, *Triturus vulgaris*, to further exposure to males. *British Journal of Herpetology* 6, 414–415.

Verrell, P.A. (1986) Male discrimination of larger, more fecund females in the smooth newt, *Triturus vulgaris*. *Journal of Herpetology* 20, 412–418.

Verrell, P.A. (1987a) The directionality of migrations of amphibians to and from a pond in southern England, with particular reference to the smooth newt, *Triturus vulgaris*. *Amphibia–Reptilia* 8, 93–100.

Verrell, P.A. (1987b) Habitat destruction and its effects on a population of smooth newts, *Triturus vulgaris*: an unfortunate field experiment. *Herpetological Journal* 1, 175–177.

Verrell, P.A. (1989) Male mate choice for fecund females in a plethodontid salamander. *Animal Behaviour* 38, 1086–1088.

Verrell, P.A. (1994) Males may choose larger females as mates in the salamander *Desmognathus fuscus*. *Animal Behaviour* 47, 1465–1467.

Verrell, P.A. and Brown, L.E. (1993) Competition among females for mates in a species with male parental care, the midwife toad *Alytes obstetricans*. *Ethology* 93, 247–257.

Verrell, P.A. and Halliday, T.R. (1985) Autumnal migration and aquatic overwintering in the common frog, *Rana temporaria*. *British Journal of Herpetology* 6, 433–434.

Villa, J. (1979) Two fungi lethal to frog eggs in Central America. *Copeia* 1979, 650–655.

Wake, D. (1991) Declining amphibian populations. *Science* 253, 860.

Waldman, B. and Adler, K. (1979) Toad tadpoles associate preferentially with siblings. *Nature* 282, 611–613.

Waldman, B., Rice, J.E. and Honeycutt, R.L. (1992) Kin recognition and incest avoidance in toads. *American Zoologist* 32, 18–30.

Wallis, G.P. and Arntzen, J.W. (1989) Mitochondrial DNA variation in the crested newt superspecies: limited cytoplasmic gene flow among species. *Evolution* 43, 88–104.

Walls, S.C. (1990) Interference competition in postmetamorphic salamanders: interspecific differences in aggression by coexisting species. *Ecology* 71, 307–314.

Waringer-Loschenkohl, A. (1988) An experimental study of microhabitat selection and microhabitat shifts in European tadpoles. *Amphibia–Reptilia* 9, 219–236.

Warner, S.C., Travis, J. and Dunson, W.A. (1993) Effect of pH variation on interspecific competition between two species of hylid tadpoles. *Ecology* 74, 183–194.

Wassersug, R.J. (1973) Aspects of social behaviour in anuran larvae, in *Evolutionary Biology of the Anurans. Contemporary research on major problems* (ed. J.L. Vial). University of Missouri Press, Columbia.

Wassersug, R.J. (1975) The adaptive significance of the tadpole stage with comments on the maintenance of complex life cycles in anurans. *American Zoologist* 15, 405–417.

Webb, N.R. (1989) Studies on the invertebrate fauna of fragmented heathland in Dorset, UK, and the implications for conservation. *Biological Conservation* 47, 155.

Webb, N.R. (1990) Changes on the heathlands of Dorset, England, between 1978 and 1987. *Biological Conservation* 51, 273–286.

Webb, N.R. and Haskins, L.E. (1980) An ecological survey of heathlands in the Poole Basin, Dorset, England in 1978. *Biological Conservation* 17, 281–296.

Wells, K.D. (1977) The social behaviour of anuran amphibians. *Animal Behaviour* 25, 666–693.

Werner, E.E. (1986) Amphibian metamorphosis: growth rate, predation risk and the optimal size at transformation. *American Naturalist* 128, 319–341.

Werner, E.E. (1991) Nonlethal effects of a predator on competitive interactions between two anuran larvae. *Ecology* 72, 1709–1720.

Werner, E.E. and McPeek, M.A. (1994) Direct and indirect effects of predators on two anuran species along an environmental gradient. *Ecology* 75, 1368–1382.

Wijnands, H.E.J. (1982) Electrophoresis of proteins in small blood samples from live frogs, in 1mm thick polyacrylamide gels. *Amphibia–Reptilia* 2, 315–319.

Wilbur, H.M. (1972) Competition, predation, and the structure of the *Ambystoma–Rana sylvatica* community. *Ecology* 53, 3–21.

Wilbur, H.M. (1980) Complex life cycles. *Annual Review of Ecology and Systematics* 11, 67–93.

Wilbur, H.M. (1982) Competition between tadpoles of *Hyla femoralis* and *Hyla gratiosa* in laboratory experiments. *Ecology* 63, 278–282.

Wilbur, H.M. (1987) Regulation of structure in complex systems: experimental temporary pond communities. *Ecology* 68, 1437–1452.

Wilbur, H.M. (1990) Coping with chaos: toads in ephemeral ponds. *Trends in Ecology and Evolution* 5, 37.

Wilbur, H.M. and Alford, R.A. (1985) Priority effects in experimental pond communities: responses of *Hyla* to *Bufo* and *Rana*. *Ecology* 66, 1106–1114.

Wilbur, H.M. and Collins, J.P. (1973) Ecological aspects of amphibian metamorphosis. *Science* 182, 1305–1314.

Wilbur, H.M. and Fauth, J.E. (1990) Experimental aquatic food webs: interactions between two predators and two prey. *American Naturalist* 135, 176–204.

Wilkin, P.J. and Scofield, A.M. (1990) The use of a serological technique to examine host selection in a natural population of the medicinal leech, *Hirudo medicinalis*. *Freshwater Biology* 23, 165–169.

Williams, G. and Hall, M. (1987) The loss of coastal grazing marshes in south and east England, with special reference to east Essex, England. *Biological Conservation* 39, 243–253.

Williams, J.G.K., Kubelik, A.R., Livak, K.J. *et al.* (1990) DNA polymorphisms amplified by arbitrary primers are useful as genetic markers. *Nucleic Acids Research* 18, 6531–6535.

Wong, A. L.-C. and Beebee, T.J.C. (1994) Identification of a unicellular, non-pigmented alga that mediates growth inhibition in anuran tadpoles: a new species of the genus *Prototheca* (Chlorophyceae: Chlorococcales). *Hydrobiologia* 277, 85–96.

Wong, A.L.-C., Beebee, T.J.C. and Griffiths, R.A. (1994) Factors affecting the distribution and abundance of an unpigmented heterotrophic alga *Prototheca richardsi*. *Freshwater Biology* 32, 33–38.

Wyman, R.L. and Hawksley-Lescault, D.S. (1987) Soil acidity affects distribution, behaviour and physiology of the salamander *Plethodon cinereus*. *Ecology* 68, 819–1827.

Yalden, D.W. (1980) An alternative explanation of the distribution of the rare herptiles in Britain. *British Journal of Herpetology* 6, 37–40.

Yalden, D.W. (1986) The distribution of newts, *Triturus* spp., in the Peak District, England. *Herpetological Journal* 1, 97–101.

Zuiderwijk, A. (1990) Sexual strategies in the newts *Triturus cristatus* and *Triturus marmoratus*. *Bijdragen tot der Dierkunde* 60, 51–64.

Zuiderwijk, A. and Bouton, N. (1987) On competition in the genus *Triturus*. *Proceedings of the 4th Ordinary General Meeting of the Societas Europaea Herpetologica*, 453–458.

Zuiderwijk, A. and Sparreboom, M. (1986) Territorial behaviour in crested newt *Triturus cristatus* and marbled newt *T. marmoratus* (Amphibia: Urodela). *Bijdragen tot de Dierkunde* 56, 205–213.

Appendix

COMMON/VERNACULAR AND LATIN NAMES OF SPECIES REFERRED TO IN TEXT

African bullfrog	*Pyxicephalus adspersus*
Agile frog	*Rana dalmatina*
Alpine newt	*Triturus alpestris*
Alpine salamander	*Salamandra atra*
American toad	*Bufo americanus*
Amphiuma salamanders	*Amphiuma* species
Arrow-poison frogs	*Dendrobates, Phyllobates*
Axolotl	*Ambystoma mexicanum*
Banded newt	*Triturus vittatus*
Barking tree frog	*Hyla gratiosa*
Black-bellied salamander	*Desmognathus quadramaculatus*
Blue-spotted salamander	*Ambystoma laterale*
Bullfrog	*Rana catesbiana*
California tree frog	*Hyla cadaverina*
Cascades frog	*Rana cascadae*
Cave salamander (European)	*Speleomantes (=Hydromantes) ambrosii*
(North American)	*Eurycea lucifuga*
Cherokee salamander	*Desmognathus aeneus*
Clawed frogs/toads	*Xenopus* species
Common frog (European)	*Rana temporaria*
Common newt	*Triturus vulgaris*
Common toad	*Bufo bufo*
Crab-eating frog	*Rana cancrivora*
Crested newt (superspecies)	*Triturus carnifex/cristatus/ dobrogicus/karelini*
Cricket frog	*Acris crepitans*
Dakota toad	*Bufo hemiophrys*
Darwin's frog	*Rhinoderma darwini*
Eastern newt, *see* Red-spotted newt	
Edible frog	*Rana kl. esculenta*
European tree frog	*Hyla arborea*

Fire-bellied toad	*Bombina bombina*
Fire salamander	*Salamandra salamandra*
Gastric brooding frogs	*Rheobatrachus* species
Giant salamanders	*Andrias* species
Golden toad	*Bufo periglenes*
Golden-striped salamander	*Chioglossa lusitanica*
Goliath frog	*Conraua goliath*
Gopher frog	*Rana areolata*
Greater gray tree frog	*Hyla versicolor*
Green frog	*Rana clamitans*
Green toad	*Bufo viridis*
Green tree frog	*Hyla cinerea*
Gulf coast toad	*Bufo valliceps*
Horned toads	*Ceratophrys* species
Houston toad	*Bufo houstonensis*
Indian bullfrog	*Rana tigrina*
Jefferson's salamander	*Ambystoma jeffersonianum*
Jordan's salamander	*Plethodon jordani*
Larch mountain salamander	*Plethodon larselli*
Leopard frog	*Rana pipiens*
Lesser gray tree frog	*Hyla chrysocelis*
Marbled newt	*Triturus marmoratus*
Marbled salamander	*Ambystoma opacum*
Marine toad	*Bufo marinus*
Marsh frog	*Rana ridibunda/perezi*
Marsupial frogs	*Gastrotheca* species
Midwife toad	*Alytes obstetricans*
– Majorcan	*Alytes muletensis*
Mink frog	*Rana septentrionalis*
Moor frog	*Rana arvalis*
Mountain salamander	*Desmognathus ochrophaeus*
Mudpuppy	*Necturus maculosus*
Natterjack toad	*Bufo calamita*
Northern chorus frog	*Pseudacris triseriata*
Northern dusky salamander	*Desmognathus fuscus*
Olm	*Proteus anguineus*
Ornate chorus frog	*Pseudacris ornata*
Pacific tree frog	*Hyla regilla*
Painted frogs	*Discoglossus* species
Painted reed frog	*Hyperolius marmoratus*
Palmate newt	*Triturus helveticus*
Parsley frog	*Pelodytes punctatus*

Pickerel frog	*Rana palustris*
Pigmy salamander	*Desmognathus wrighti*
Pine Barrens tree frog	*Hyla andersoni*
Pine woods tree frog	*Hyla femoralis*
Pool frog	*Rana lessonae*
Pyrenean brook salamander	*Euproctus asper*
Raucous toad	*Bufo rangeri*
Red-backed salamander	*Plethodon cinereus*
Red-bellied newt	*Taricha rivularis*
Red Hills salamander	*Phaeognathus hubrichti*
Red-legged frog	*Rana aurora*
Red-spotted newt	*Notophthalmus viridiscens*
Seal salamander	*Desmognathus monticola*
Sharp-nosed torrent frog	*Taudactylus acutirostris*
Sharp-ribbed salamander	*Pleurodeles waltl*
Siren salamanders	*Siren* spp
Slimy salamander	*Plethodon glutinosus*
Small-mouth salamander	*Ambystoma texanum*
Smooth newt, *see* Common newt	
Southern day frog	*Taudactylus diurnus*
Southern leopard frog	*Rana sphenocephala*
Southern toad	*Bufo terrestris*
Spadefoot toads	
Couch's	*Scaphiopus couchii*
Eastern (N. America)	*Scaphiopus holbrooki*
Italian	*Pelobates fuscus insubricus*
Plains	*Scaphiopus (=Spea) bombifrons*
Western (Europe)	*Pelobates cultripes*
Spotted frog	*Rana pretiosa*
Spotted salamander	*Ambystoma maculatum*
Spotted tree frog	*Limnodynastes spenceri*
Spring peeper	*Pseudacris (=Hyla) crucifer*
Surinam toad	*Pipa pipa*
Tailed frog	*Ascaphus truei*
Talpid salamander	*Ambystoma talpoideum*
Tarahumara frog	*Rana tarahumarae*
Texas salamander	*Eurycea neotenes*
Tiger salamander	*Ambystoma tigrinum*
Tremblay's salamander	*Ambystoma tremblayi*
Valdina farms salamander	*Eurycea troglodytes*
Western toad	*Bufo boreas*
White's tree frog	*Litoria caerulea*

Wood frog	*Rana sylvatica*
Woodhouse's toad	*Bufo woodhousei*
Yellow-bellied toad	*Bombina variegata*
Yellow-legged frog	*Rana muscosa*
Yosemite toad	Bufo canorus

Index